These safety symbols are used in laboratory and field investigations in this bo[ok]
ing of each symbol and refer to this page often. *Remember to wash your han[ds]*

PROTECTIVE EQUIPMENT Do not begin any lab without the prop[er]

 GOGGLES Proper eye protection must be worn when performing or observing science activities that involve items or conditions as listed below.

 APRON Wear an approved apron when using substances that could stain, wet, or destroy cloth.

 SOAP W[ash with] soap and water [before] removing goggles and after all lab activities.

 [GLOVES Wear gloves when working with] materials, chemicals, animals, or materials that can stain or irritate hands.

LABORATORY HAZARDS

Symbols	Potential Hazards	Precaution	Response
DISPOSAL	contamination of classroom or environment due to improper disposal of materials such as chemicals and live specimens	• DO NOT dispose of hazardous materials in the sink or trash can. • Dispose of wastes as directed by your teacher.	• If hazardous materials are disposed of improperly, notify your teacher immediately.
EXTREME TEMPERATURE	skin burns due to extremely hot or cold materials such as hot glass, liquids, or metals; liquid nitrogen; dry ice	• Use proper protective equipment, such as hot mitts and/or tongs, when handling objects with extreme temperatures.	• If injury occurs, notify your teacher immediately.
SHARP OBJECTS	punctures or cuts from sharp objects such as razor blades, pins, scalpels, and broken glass	• Handle glassware carefully to avoid breakage. • Walk with sharp objects pointed downward, away from you and others.	• If broken glass or injury occurs, notify your teacher immediately.
ELECTRICAL	electric shock or skin burn due to improper grounding, short circuits, liquid spills, or exposed wires	• Check condition of wires and apparatus for fraying or uninsulated wires, and broken or cracked equipment. • Use only GFCI-protected outlets	• DO NOT attempt to fix electrical problems. Notify your teacher immediately.
CHEMICAL	skin irritation or burns, breathing difficulty, and/or poisoning due to touching, swallowing, or inhalation of chemicals such as acids, bases, bleach, metal compounds, iodine, poinsettias, pollen, ammonia, acetone, nail polish remover, heated chemicals, mothballs, and any other chemicals labeled or known to be dangerous	• Wear proper protective equipment such as goggles, apron, and gloves when using chemicals. • Ensure proper room ventilation or use a fume hood when using materials that produce fumes. • NEVER smell fumes directly. • NEVER taste or eat any material in the laboratory.	• If contact occurs, immediately flush affected area with water and notify your teacher. • If a spill occurs, leave the area immediately and notify your teacher.
FLAMMABLE	unexpected fire due to liquids or gases that ignite easily such as rubbing alcohol	• Avoid open flames, sparks, or heat when flammable liquids are present.	• If a fire occurs, leave the area immediately and notify your teacher.
OPEN FLAME	burns or fire due to open flame from matches, Bunsen burners, or burning materials	• Tie back loose hair and clothing. • Keep flame away from all materials. • Follow teacher instructions when lighting and extinguishing flames. • Use proper protection, such as hot mitts or tongs, when handling hot objects.	• If a fire occurs, leave the area immediately and notify your teacher.
ANIMAL SAFETY	injury to or from laboratory animals	• Wear proper protective equipment such as gloves, apron, and goggles when working with animals. • Wash hands after handling animals.	• If injury occurs, notify your teacher immediately.
BIOLOGICAL	infection or adverse reaction due to contact with organisms such as bacteria, fungi, and biological materials such as blood, animal or plant materials	• Wear proper protective equipment such as gloves, goggles, and apron when working with biological materials. • Avoid skin contact with an organism or any part of the organism. • Wash hands after handling organisms.	• If contact occurs, wash the affected area and notify your teacher immediately.
FUME	breathing difficulties from inhalation of fumes from substances such as ammonia, acetone, nail polish remover, heated chemicals, and mothballs	• Wear goggles, apron, and gloves. • Ensure proper room ventilation or use a fume hood when using substances that produce fumes. • NEVER smell fumes directly.	• If a spill occurs, leave area and notify your teacher immediately.
IRRITANT	irritation of skin, mucous membranes, or respiratory tract due to materials such as acids, bases, bleach, pollen, mothballs, steel wool, and potassium permanganate	• Wear goggles, apron, and gloves. • Wear a dust mask to protect against fine particles.	• If skin contact occurs, immediately flush the affected area with water and notify your teacher.
RADIOACTIVE	excessive exposure from alpha, beta, and gamma particles	• Remove gloves and wash hands with soap and water before removing remainder of protective equipment.	• If cracks or holes are found in the container, notify your teacher immediately.

Your online portal to everything you need

connectED.mcgraw-hill.com

Look for these icons to access exciting digital resources

- Video
- Audio
- Review
- Inquiry
- WebQuest
- Assessment
- Concepts in Motion

McGraw Hill Education

WEATHER AND CLIMATE

iSCIENCE

Glencoe

Geode
This is a cross section of geode, a type of rock. The outside of a geode is generally limestone, but the inside contains mineral crystals. Crystals only partially fill this geode, but other geodes are filled completely with crystals.

The McGraw·Hill Companies

 Education

Copyright © 2012 The McGraw-Hill Companies, Inc. All rights reserved. No part of this publication may be reproduced or distributed in any form or by any means, or stored in a database or retrieval system, without the prior written consent of The McGraw-Hill Companies, Inc., including, but not limited to, network storage or transmission, or broadcast for distance learning.

Send all inquiries to:
McGraw-Hill Education
8787 Orion Place
Columbus, OH 43240-4027

ISBN: 978-0-07-888010-0
MHID: 0-07-888010-6

Printed in the United States of America.

3 4 5 6 7 8 9 10 DOW 15 14 13 12

Authors and Contributors

Authors

American Museum of Natural History
New York, NY

Michelle Anderson, MS
Lecturer
The Ohio State University
Columbus, OH

Juli Berwald, PhD
Science Writer
Austin, TX

John F. Bolzan, PhD
Science Writer
Columbus, OH

Rachel Clark, MS
Science Writer
Moscow, ID

Patricia Craig, MS
Science Writer
Bozeman, MT

Randall Frost, PhD
Science Writer
Pleasanton, CA

Lisa S. Gardiner, PhD
Science Writer
Denver, CO

Jennifer Gonya, PhD
The Ohio State University
Columbus, OH

Mary Ann Grobbel, MD
Science Writer
Grand Rapids, MI

Whitney Crispen Hagins, MA, MAT
Biology Teacher
Lexington High School
Lexington, MA

Carole Holmberg, BS
Planetarium Director
Calusa Nature Center and Planetarium, Inc.
Fort Myers, FL

Tina C. Hopper
Science Writer
Rockwall, TX

Jonathan D. W. Kahl, PhD
Professor of Atmospheric Science
University of Wisconsin-Milwaukee
Milwaukee, WI

Nanette Kalis
Science Writer
Athens, OH

S. Page Keeley, MEd
Maine Mathematics and Science Alliance
Augusta, ME

Cindy Klevickis, PhD
Professor of Integrated Science and Technology
James Madison University
Harrisonburg, VA

Kimberly Fekany Lee, PhD
Science Writer
La Grange, IL

Michael Manga, PhD
Professor
University of California, Berkeley
Berkeley, CA

Devi Ried Mathieu
Science Writer
Sebastopol, CA

Elizabeth A. Nagy-Shadman, PhD
Geology Professor
Pasadena City College
Pasadena, CA

William D. Rogers, DA
Professor of Biology
Ball State University
Muncie, IN

Donna L. Ross, PhD
Associate Professor
San Diego State University
San Diego, CA

Marion B. Sewer, PhD
Assistant Professor
School of Biology
Georgia Institute of Technology
Atlanta, GA

Julia Meyer Sheets, PhD
Lecturer
School of Earth Sciences
The Ohio State University
Columbus, OH

Michael J. Singer, PhD
Professor of Soil Science
Department of Land, Air and Water Resources
University of California
Davis, CA

Karen S. Sottosanti, MA
Science Writer
Pickerington, Ohio

Paul K. Strode, PhD
I.B. Biology Teacher
Fairview High School
Boulder, CO

Jan M. Vermilye, PhD
Research Geologist
Seismo-Tectonic Reservoir Monitoring (STRM)
Boulder, CO

Judith A. Yero, MA
Director
Teacher's Mind Resources
Hamilton, MT

Dinah Zike, MEd
Author, Consultant,
Inventor of Foldables
Dinah Zike Academy;
Dinah-Might Adventures, LP
San Antonio, TX

Margaret Zorn, MS
Science Writer
Yorktown, VA

Consulting Authors

Alton L. Biggs
Biggs Educational Consulting
Commerce, TX

Ralph M. Feather, Jr., PhD
Assistant Professor
Department of Educational
Studies and Secondary
Education
Bloomsburg University
Bloomsburg, PA

Douglas Fisher, PhD
Professor of Teacher Education
San Diego State University
San Diego, CA

Edward P. Ortleb
Science/Safety Consultant
St. Louis, MO

Series Consultants

Science

Solomon Bililign, PhD
Professor
Department of Physics
North Carolina Agricultural
and Technical State University
Greensboro, NC

John Choinski
Professor
Department of Biology
University of Central Arkansas
Conway, AR

Anastasia Chopelas, PhD
Research Professor
Department of Earth and
Space Sciences
UCLA
Los Angeles, CA

David T. Crowther, PhD
Professor of Science Education
University of Nevada, Reno
Reno, NV

A. John Gatz
Professor of Zoology
Ohio Wesleyan University
Delaware, OH

Sarah Gille, PhD
Professor
University of California
San Diego
La Jolla, CA

David G. Haase, PhD
Professor of Physics
North Carolina State
University
Raleigh, NC

Janet S. Herman, PhD
Professor
Department of Environmental
Sciences
University of Virginia
Charlottesville, VA

David T. Ho, PhD
Associate Professor
Department of Oceanography
University of Hawaii
Honolulu, HI

Ruth Howes, PhD
Professor of Physics
Marquette University
Milwaukee, WI

Jose Miguel Hurtado, Jr., PhD
Associate Professor
Department of Geological
Sciences
University of Texas at El Paso
El Paso, TX

Monika Kress, PhD
Assistant Professor
San Jose State University
San Jose, CA

Mark E. Lee, PhD
Associate Chair & Assistant
Professor
Department of Biology
Spelman College
Atlanta, GA

Linda Lundgren
Science writer
Lakewood, CO

Series Consultants, continued

Keith O. Mann, PhD
Ohio Wesleyan University
Delaware, OH

Charles W. McLaughlin, PhD
Adjunct Professor of Chemistry
Montana State University
Bozeman, MT

Katharina Pahnke, PhD
Research Professor
Department of Geology and Geophysics
University of Hawaii
Honolulu, HI

Jesús Pando, PhD
Associate Professor
DePaul University
Chicago, IL

Hay-Oak Park, PhD
Associate Professor
Department of Molecular Genetics
Ohio State University
Columbus, OH

David A. Rubin, PhD
Associate Professor of Physiology
School of Biological Sciences
Illinois State University
Normal, IL

Toni D. Sauncy
Assistant Professor of Physics
Department of Physics
Angelo State University
San Angelo, TX

Malathi Srivatsan, PhD
Associate Professor of Neurobiology
College of Sciences and Mathematics
Arkansas State University
Jonesboro, AR

Cheryl Wistrom, PhD
Associate Professor of Chemistry
Saint Joseph's College
Rensselaer, IN

Reading

ReLeah Cossett Lent
Author/Educational Consultant
Blue Ridge, GA

Math

Vik Hovsepian
Professor of Mathematics
Rio Hondo College
Whittier, CA

Series Reviewers

Thad Boggs
Mandarin High School
Jacksonville, FL

Catherine Butcher
Webster Junior High School
Minden, LA

Erin Darichuk
West Frederick Middle School
Frederick, MD

Joanne Hedrick Davis
Murphy High School
Murphy, NC

Anthony J. DiSipio, Jr.
Octorara Middle School
Atglen, PA

Adrienne Elder
Tulsa Public Schools
Tulsa, OK

Series Reviewers, continued

Carolyn Elliott
Iredell-Statesville Schools
Statesville, NC

Christine M. Jacobs
Ranger Middle School
Murphy, NC

Jason O. L. Johnson
Thurmont Middle School
Thurmont, MD

Felecia Joiner
Stony Point Ninth Grade Center
Round Rock, TX

Joseph L. Kowalski, MS
Lamar Academy
McAllen, TX

Brian McClain
Amos P. Godby High School
Tallahassee, FL

Von W. Mosser
Thurmont Middle School
Thurmont, MD

Ashlea Peterson
Heritage Intermediate Grade Center
Coweta, OK

Nicole Lenihan Rhoades
Walkersville Middle School
Walkersvillle, MD

Maria A. Rozenberg
Indian Ridge Middle School
Davie, FL

Barb Seymour
Westridge Middle School
Overland Park, KS

Ginger Shirley
Our Lady of Providence Junior-Senior High School
Clarksville, IN

Curtis Smith
Elmwood Middle School
Rogers, AR

Sheila Smith
Jackson Public School
Jackson, MS

Sabra Soileau
Moss Bluff Middle School
Lake Charles, LA

Tony Spoores
Switzerland County Middle School
Vevay, IN

Nancy A. Stearns
Switzerland County Middle School
Vevay, IN

Kari Vogel
Princeton Middle School
Princeton, MN

Alison Welch
Wm. D. Slider Middle School
El Paso, TX

Linda Workman
Parkway Northeast Middle School
Creve Coeur, MO

Teacher Advisory Board

The Teacher Advisory Board gave the authors, editorial staff, and design team feedback on the content and design of the Student Edition. They provided valuable input in the development of *Glencoe ⓘScience*.

Frances J. Baldridge
Department Chair
Ferguson Middle School
Beavercreek, OH

Jane E. M. Buckingham
Teacher
Crispus Attucks Medical
Magnet High School
Indianapolis, IN

Elizabeth Falls
Teacher
Blalack Middle School
Carrollton, TX

Nelson Farrier
Teacher
Hamlin Middle School
Springfield, OR

Michelle R. Foster
Department Chair
Wayland Union
Middle School
Wayland, MI

Rebecca Goodell
Teacher
Reedy Creek Middle School
Cary, NC

Mary Gromko
Science Supervisor K–12
Colorado Springs District 11
Colorado Springs, CO

Randy Mousley
Department Chair
Dean Ray Stucky
Middle School
Wichita, KS

David Rodriguez
Teacher
Swift Creek Middle School
Tallahassee, FL

Derek Shook
Teacher
Floyd Middle Magnet School
Montgomery, AL

Karen Stratton
Science Coordinator
Lexington School District One
Lexington, SC

Stephanie Wood
Science Curriculum Specialist,
K–12
Granite School District
Salt Lake City, UT

Online Guide

Get ConnectED
connectED.mcgraw-hill.com

ConnectED
▷ **Your Digital Science Portal**

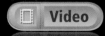  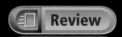 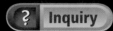 

See the science in real life through these exciting videos.

Click the link and you can listen to the text while you follow along.

Try these interactive tools to help you review the lesson concepts.

Explore concepts through hands-on and virtual labs.

These web-based challenges relate the concepts you're learning about to the latest news and research.

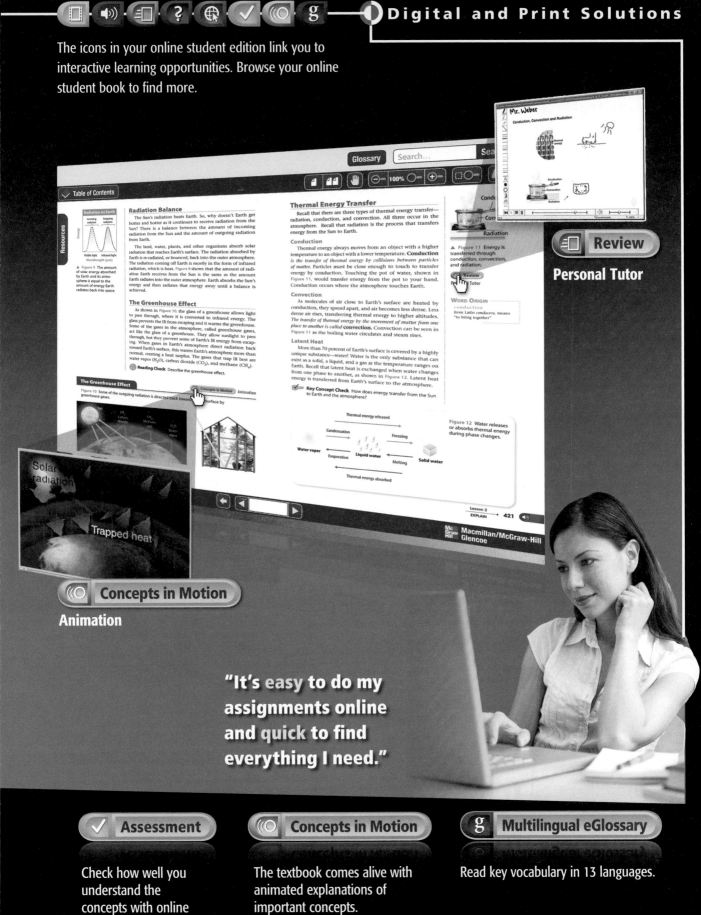

Digital and Print Solutions

The icons in your online student edition link you to interactive learning opportunities. Browse your online student book to find more.

"It's easy to do my assignments online and quick to find everything I need."

Assessment
Check how well you understand the concepts with online quizzes and practice questions.

Concepts in Motion
The textbook comes alive with animated explanations of important concepts.

Multilingual eGlossary
Read key vocabulary in 13 languages.

Treasure Hunt

Your science book has many features that will aid you in your learning. Some of these features are listed below. You can use the activity at the right to help you find these and other special features in the book.

- **THE BIG IDEA** can be found at the start of each chapter.
- The Reading Guide at the start of each lesson lists 🗝 **Key Concepts**, vocabulary terms, and online supplements to the content.
- **Connect ED** icons direct you to online resources such as animations, personal tutors, math practices, and quizzes.
- **Inquiry** Labs and Skill Practices are in each chapter.
- Your **FOLDABLES** help organize your notes.

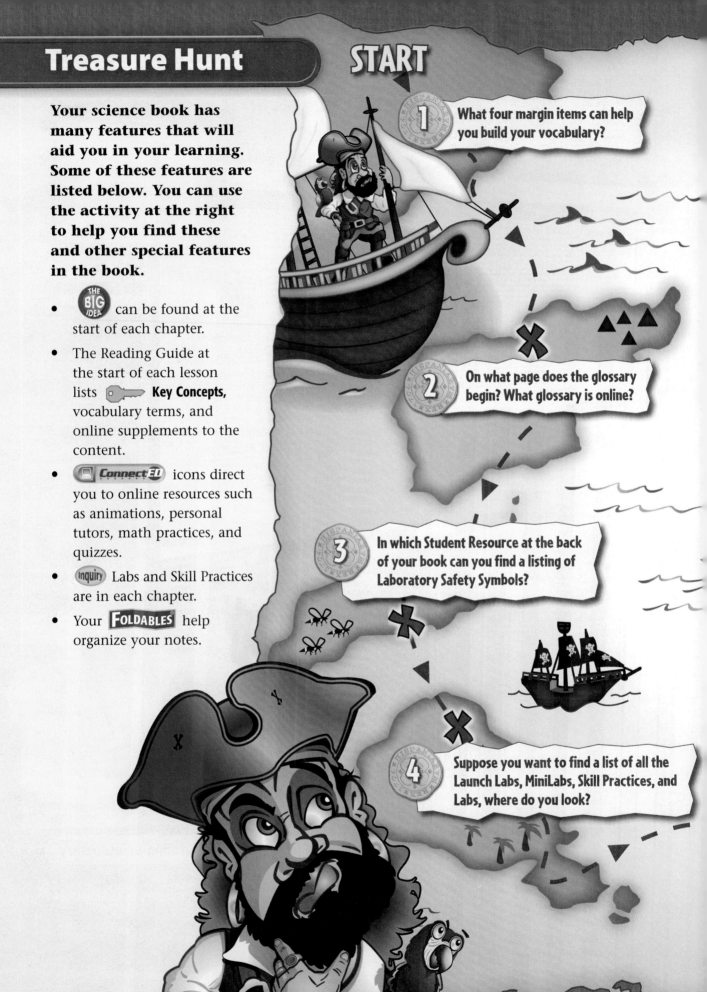

START

1. What four margin items can help you build your vocabulary?

2. On what page does the glossary begin? What glossary is online?

3. In which Student Resource at the back of your book can you find a listing of Laboratory Safety Symbols?

4. Suppose you want to find a list of all the Launch Labs, MiniLabs, Skill Practices, and Labs, where do you look?

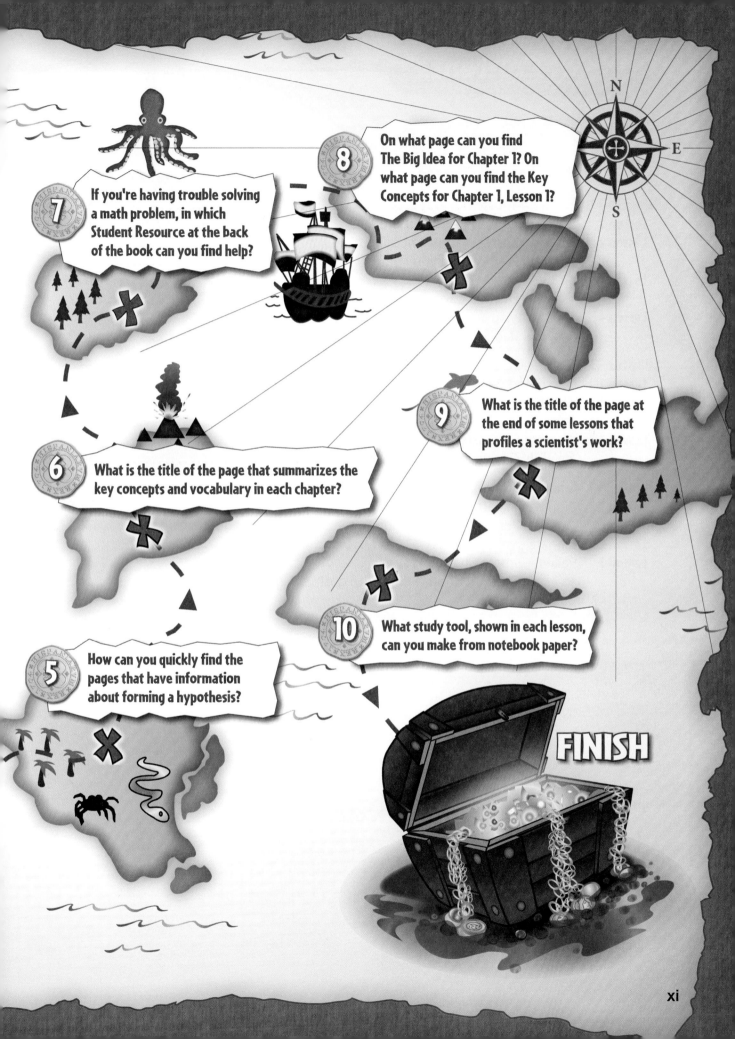

Table of Contents

Unit 3 **Weather and Climate** **402**

Chapter 12 **Earth's Atmosphere** **406**
Lesson 1 Describing Earth's Atmosphere 408
Lesson 2 Energy Transfer in the Atmosphere 417
 Skill Practice Can you conduct, convect, and radiate? 425
Lesson 3 Air Currents 426
 Skill Practice Can you model global wind patterns? 432
Lesson 4 Air Quality 433
 Lab Radiant Energy Absorption 440

Chapter 13 **Weather** **448**
Lesson 1 Describing Weather 450
Lesson 2 Weather Patterns 458
 Skill Practice Why does the weather change? 469
Lesson 3 Weather Forecasts 470
 Lab Can you predict the weather? 476

Chapter 14 **Climate** **484**
Lesson 1 Climates of Earth 486
 Skill Practice Can reflection of the Sun's rays change the climate? 494
Lesson 2 Climate Cycles 495
Lesson 3 Recent Climate Change 504
 Lab The greenhouse effect is a gas! 512

Table of Contents

Student Resources

Science Skill Handbook **SR-2**
  Scientific Methods SR-2
  Safety Symbols SR-11
  Safety in the Science Laboratory SR-12

Math Skill Handbook **SR-14**
  Math Review SR-14
  Science Application SR-24

Foldables Handbook **SR-29**

Reference Handbook **SR-40**
  Periodic Table of the Elements SR-40
  Topographic Map Symbols SR-42
  Rocks SR-43
  Minerals SR-44
  Weather Map Symbols SR-46

Glossary **G-2**

Index **I-2**

Credits **C-2**

Inquiry

Launch Labs

12-1	Where does air apply pressure?	409
12-2	What happens to air as it warms?	418
12-3	Why does air move?	427
12-4	How does acid rain form?	434
13-1	Can you make clouds in a bag?	451
13-2	How can temperature affect pressure?	459
13-3	Can you understand the weather report?	471
14-1	How do climates compare?	487
14-2	How does Earth's tilted axis affect climate?	496
14-3	What changes climates?	505

MiniLabs

12-1	Why does the furniture get dusty?	410
12-2	Can you identify a temperature inversion?	423
12-3	Can you model the Coriolis effect?	429
12-4	Can being out in fresh air be harmful to your health?	437
13-1	When will dew form?	453
13-2	How can you observe air pressure?	461
13-3	How is weather represented on a map?	474
14-1	Where are microclimates found?	492
14-2	How do climates vary?	501
14-3	How much CO_2 do vehicles emit?	509

Inquiry

Inquiry Skill Practice

- **12-2** Can you conduct, convect, and radiate? 425
- **12-3** Can you model global wind patterns? 432
- **13-2** Why does the weather change? 469
- **14-1** Can reflection of the Sun's rays change the climate? 494

Inquiry Labs

- **12-4** Radiant Energy Absorption 440
- **13-3** Can you predict the weather? 476
- **14-3** The greenhouse effect is a gas! 512

Features

Science & Society

- **13-1** Is there a link between hurricanes and global warming? 457

Careers in Science

- **12-1** A Crack in Earth's Shield 416
- **14-2** Frozen in Time 503

Unit 3
Weather and Climate

1441
Prince Munjong of Korea invents the first rain gauge to gather and measure the amount of liquid precipitation over a period of time.

1450
The first anemometer, a tool to measure wind speed, is developed by Leone Battista Alberti.

1643
Italian physicist Evangelista Torricelli invents the barometer to measure pressure in the air. This tool improves meteorology, which relied on simple sky observations.

1714
German physicist Daniel Fahrenheit develops the mercury thermometer, making it possible to measure temperature.

1752
Swedish astronomer Andres Celsius proposes a centigrade temperature scale where 0° is the freezing point of water and 100° is the boiling point of water.

1806 Francis Beaufort creates a system for naming wind speeds and aptly names it the Beaufort Wind Force Scale. This scale is used mainly to classify sea conditions.

1960 TIROS 1, the world's first weather satellite, is sent into space equipped with a TV camera.

1964 The U.S. National Severe Storms Laboratory begins experimenting with the use of Doppler radar for weather-monitoring purposes.

2006 Meteorologists hold 8,800 jobs in the United States alone. These scientists work in government and private agencies, in research services, on radio and television stations, and in education.

Inquiry

Visit ConnectED for this unit's STEM activity.

Unit 3 Nature of SCIENCE

Models

In 2004 over 200,000 people died as a tsunami swept across the Indian Ocean, shown in **Figure 1**. How can scientists predict future tsunamis to help save lives? Researchers around the world have developed different models to study tsunami waves and their effects. A **model** is a representation of an object, a process, an event, or a system that is similar to the physical object or idea being explained. Scientists use models to study something that is too big or too small, happens too quickly or too slowly, or is too dangerous or too expensive to study directly.

Models of tsunamis help predict how future tsunamis might impact land. Information from these models can help save ecosystems, buildings, and lives.

▲ **Figure 1** A massive wave approaches the shore in the 2004 Indian Ocean tsunami.

Types of Models

Mathematical Models and Computer Simulations

A mathematical model represents an event, a process, or a system, using equations. A mathematical model can be one equation, for example: speed=distance/time. Or, they can be several hundred equations, such as those used to calculate tsunami effects.

A computer simulation is a model that combines many mathematical models. Computer simulations allow the user to easily change variables. Simulations often show a change over time or a sequence of events. Computer programs that include animations and graphics are used to visually display mathematical models.

Researchers from Texas A&M University constructed a tsunami simulation using many mathematical models of Seaside, Oregon, as shown in **Figure 2**. Simulations that use equations to model the force of waves hitting buildings are displayed on a computer screen. Researchers change variables, such as the size, the force, or the shape of tsunami waves, to determine how Seaside might be damaged by a tsunami.

Figure 2 This series of images is from an animated simulation model of a tsunami approaching Seaside, Oregon. ▼

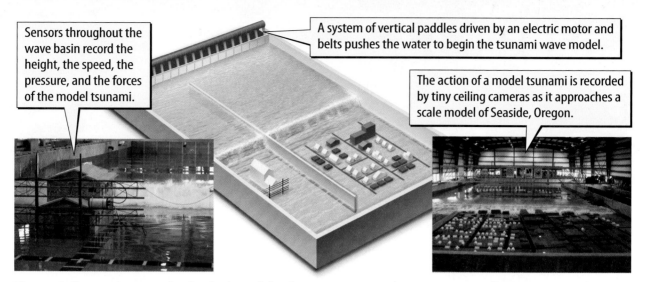

Figure 3 Researchers study physical models of tsunamis to predict a tsunami's effects.

Physical Models

A physical model is a model that you can see and touch. It shows how parts relate to one another, how something is built, or how complex objects work. Scientists at Oregon State University built physical scale models of Seaside, Oregon, as shown in **Figure 3.** They placed the model at the end of a long wave tank. Sensors in the wave tank and on the model buildings measure and record velocities, forces, and turbulence created by a model tsunami wave. Scientists use these measurements to predict the effects of a tsunami on a coastal town, and to make recommendations for saving lives and preventing damage.

Conceptual Models

Images that represent a process or relationships among ideas are conceptual models. The conceptual model below shows that the United States has a three-part plan for minimizing the effects of tsunamis. Hazard assessment involves identification of areas that are in high risk of tsunamis. Response involves education and public safety. Warning includes a system of sensors that detect the approach of a tsunami.

Inquiry MiniLab

30 minutes

How can you model a tsunami?

What tsunami behaviors can you observe in your own model of a wave tank?

1. Read and complete a lab safety form.
2. Pour **sand** into a **glass pan,** creating a slope from one end of the pan to the other. Fill the pan with water. Place a **cork** in the center of the pan. Draw your setup in your Science Journal.
3. Use a **dowel** to create a wave at the deep end of the pan. Record your observations.
4. Place several **common objects** on end at the shallow end of the pan. Record your observations of the behaviors of the cork and the different objects when you create a wave.

Analyze and Conclude

1. **Describe** What do the different parts of your physical model represent?
2. **Explain** What are some limitations of your physical model?

Models • 405

Chapter 12

Earth's Atmosphere

 How does Earth's atmosphere affect life on Earth?

Inquiry What's in the atmosphere?

Earth's atmosphere is made up of gases and small amounts of liquid and solid particles. Earth's atmosphere surrounds and sustains life.

- What type of particles make up clouds in the atmosphere?
- How do conditions in the atmosphere change as height above sea level increases?
- How does Earth's atmosphere affect life on Earth?

Get Ready to Read

What do you think?
Before you read, decide if you agree or disagree with each of these statements. As you read this chapter, see if you change your mind about any of the statements.

1. Air is empty space.
2. Earth's atmosphere is important to living organisms.
3. All the energy from the Sun reaches Earth's surface.
4. Earth emits energy back into the atmosphere.
5. Uneven heating in different parts of the atmosphere creates air circulation patterns.
6. Warm air sinks and cold air rises.
7. If no humans lived on Earth, there would be no air pollution.
8. Pollution levels in the air are not measured or monitored.

ConnectED Your one-stop online resource

connectED.mcgraw-hill.com

- Video
- WebQuest
- Audio
- Assessment
- Review
- Concepts in Motion
- Inquiry
- Multilingual eGlossary

Lesson 1

Describing Earth's Atmosphere

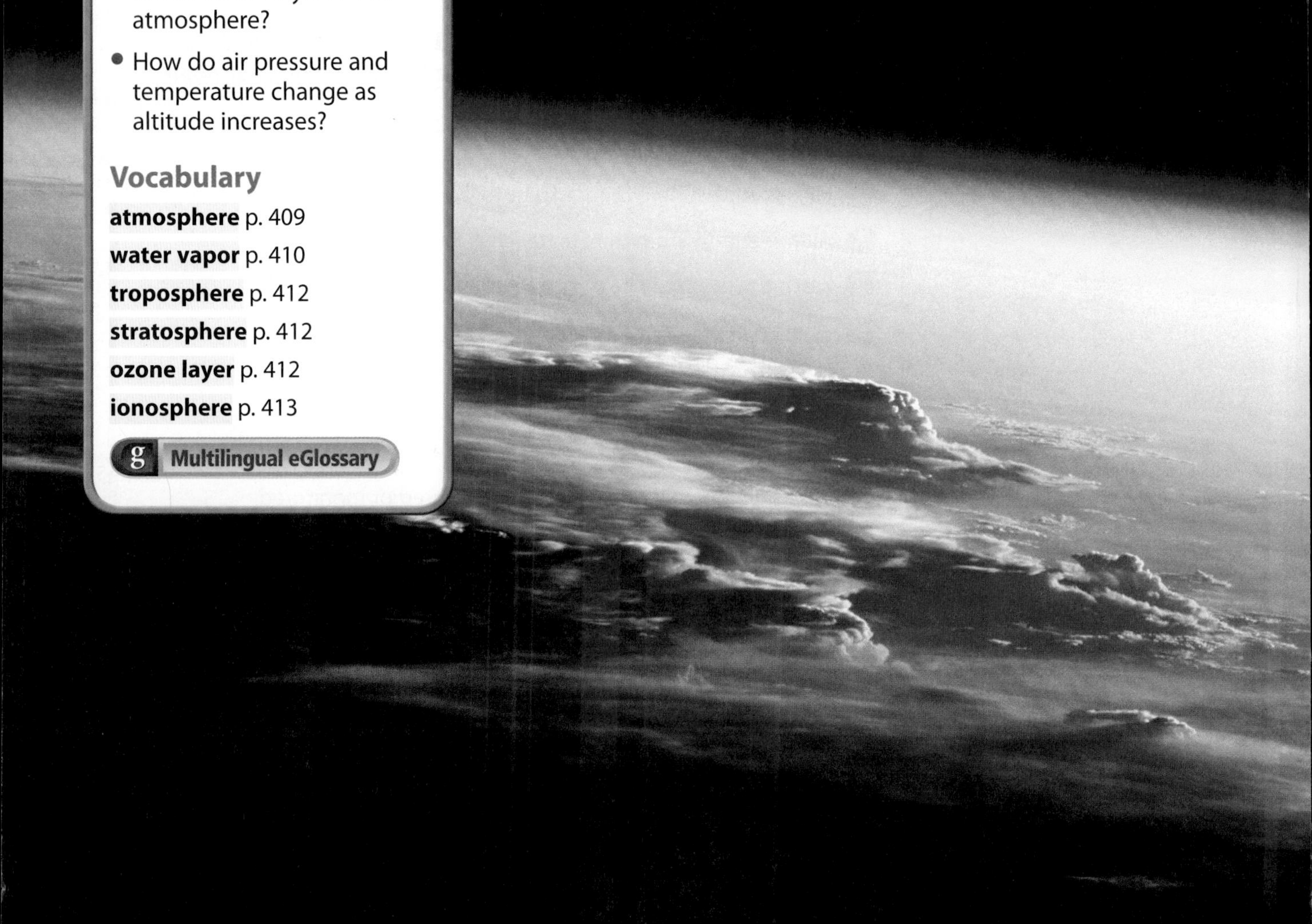

Reading Guide

Key Concepts
ESSENTIAL QUESTIONS

- How did Earth's atmosphere form?
- What is Earth's atmosphere made of?
- What are the layers of the atmosphere?
- How do air pressure and temperature change as altitude increases?

Vocabulary

atmosphere p. 409
water vapor p. 410
troposphere p. 412
stratosphere p. 412
ozone layer p. 412
ionosphere p. 413

Multilingual eGlossary

Inquiry: Why is the atmosphere important?

What would Earth be like without its atmosphere? Earth's surface would be scarred with craters created from the impact of meteorites. Earth would experience extreme daytime-to-nighttime temperature changes. How would changes in the atmosphere affect life? What effect would atmospheric changes have on weather and climate?

Launch Lab

20 minutes

Where does air apply pressure?

With the exception of Mercury, most planets in the solar system have some type of atmosphere. However, Earth's atmosphere provides what the atmospheres of other planets cannot: oxygen and water. Oxygen, water vapor, and other gases make up the gaseous mixture in the atmosphere called air. In this activity, you will explore air's effect on objects on Earth's surface.

1. Read and complete a lab safety form.
2. Add **water** to a **cup** until it is two-thirds full.
3. Place a large **index card** over the opening of the cup so that it is completely covered.
4. Hold the cup over a tub or a large bowl.
5. Place one hand on the index card to hold it in place as you quickly turn the cup upside down. Remove your hand.

Think About This

1. What happened when you turned the cup over?
2. How did air play a part in your observation?
3. **Key Concept** How do you think these results might differ if you repeated the activity in a vacuum?

Importance of Earth's Atmosphere

The photo on the previous page shows Earth's atmosphere as seen from space. How would you describe the atmosphere? *The* **atmosphere** *(AT muh sfihr) is a thin layer of gases surrounding Earth.* Earth's atmosphere is hundreds of kilometers high. However, when compared to Earth's size, it is about the same relative thickness as an apple's skin to an apple.

The atmosphere contains the oxygen, carbon dioxide, and water necessary for life on Earth. Earth's atmosphere also acts like insulation on a house. It helps keep temperatures on Earth within a range in which living organisms can survive. Without it, daytime temperatures would be extremely high and nighttime temperatures would be extremely low.

The atmosphere helps protect living organisms from some of the Sun's harmful rays. It also helps protect Earth's surface from being struck by meteors. Most meteors that fall toward Earth burn up before reaching Earth's surface. Friction with the atmosphere causes them to burn. Only the very largest meteors strike Earth.

WORD ORIGIN

atmosphere
from Greek *atmos*, means "vapor"; and Latin *sphaera*, means "sphere"

Reading Check Why is Earth's atmosphere important to life on Earth?

Origins of Earth's Atmosphere

Most scientists agree that when Earth formed, it was a ball of molten rock. As Earth slowly cooled, its outer surface hardened. Erupting volcanoes emitted hot gases from Earth's interior. These gases surrounded Earth, forming its atmosphere.

Ancient Earth's atmosphere was thought to be water vapor with a little carbon dioxide (CO_2) and nitrogen. **Water vapor** *is water in its gaseous form.* This ancient atmosphere did not have enough oxygen to support life as we know it. As Earth and its atmosphere cooled, the water vapor condensed into liquid. Rain fell and then evaporated from Earth's surface repeatedly for thousands of years. Eventually, water accumulated on Earth's surface, forming oceans. Most of the original CO_2 that dissolved in rain is in rocks on the ocean floor. Today the atmosphere has more nitrogen than CO_2.

Earth's first organisms could undergo photosynthesis, which changed the atmosphere. Recall that photosynthesis uses light energy to produce sugar and oxygen from carbon dioxide and water. The organisms removed CO_2 from the atmosphere and released oxygen into it. Eventually the levels of CO_2 and oxygen supported the development of other organisms.

Key Concept Check How did Earth's present atmosphere form?

> **REVIEW VOCABULARY**
> **liquid**
> matter with a definite volume but no definite shape that can flow from one place to another

Inquiry MiniLab
20 minutes

Why does the furniture get dusty?

Have you ever noticed that furniture gets dusty? The atmosphere is one source for dirt and dust particles. Where can you find dust in your classroom?

1. Read and complete a lab safety form.
2. Choose a place in your classroom to collect a sample of dust.
3. Using a **duster,** collect dust from about a 50-cm² area.
4. Examine the duster with a **magnifying lens.** Observe any dust particles. Some might be so small that they only make the duster look gray.
5. Record your observations in your Science Journal.
6. Compare your findings with those of other members of your class.

Analyze and Conclude

1. **Analyze** how the area surrounding your collection site might have influenced how much dust you observed on the duster.
2. **Infer** the source of the dust.
3. **Key Concept** Other than gases and water droplets, predict what Earth's atmosphere might contain.

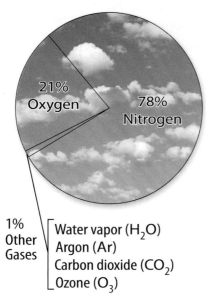

Figure 1 Oxygen and nitrogen make up most of the atmosphere, with the other gases making up only 1 percent. ▼

Visual Check What percent of the atmosphere is made up of oxygen and nitrogen?

▲ Figure 2 One way solid particles enter the atmosphere is from volcanic eruptions.

Composition of the Atmosphere

Today's atmosphere is mostly made up of invisible gases, including nitrogen, oxygen, and carbon dioxide. Some solid and liquid particles, such as ash from volcanic eruptions and water droplets, are also present.

Gases in the Atmosphere

Study **Figure 1**. Which gas is the most abundant in Earth's atmosphere? Nitrogen makes up about 78 percent of Earth's atmosphere. About 21 percent of Earth's atmosphere is oxygen. Other gases, including argon, carbon dioxide, and water vapor, make up the remaining 1 percent of the atmosphere.

The amounts of water vapor, carbon dioxide, and ozone vary. The concentration of water vapor in the atmosphere ranges from 0 to 4 percent. Carbon dioxide is 0.038 percent of the atmosphere. A small amount of ozone is at high altitudes. Ozone also occurs near Earth's surface in urban areas.

Solids and Liquids in the Atmosphere

Tiny solid particles are also in Earth's atmosphere. Many of these, such as pollen, dust, and salt, can enter the atmosphere through natural processes. **Figure 2** shows another natural source of particles in the atmosphere—ash from volcanic eruptions. Some solid particles enter the atmosphere because of human activities, such as driving vehicles that release soot.

The most common liquid particles in the atmosphere are water droplets. Although microscopic in size, water droplets are visible when they form clouds. Other atmospheric liquids include acids that result when volcanoes erupt and fossil fuels are burned. Sulfur dioxide and nitrous oxide combine with water vapor in the air and form the acids.

Key Concept Check What is Earth's atmosphere made of?

Lesson 1
EXPLAIN

Layers of Atmosphere

Concepts in Motion Animation

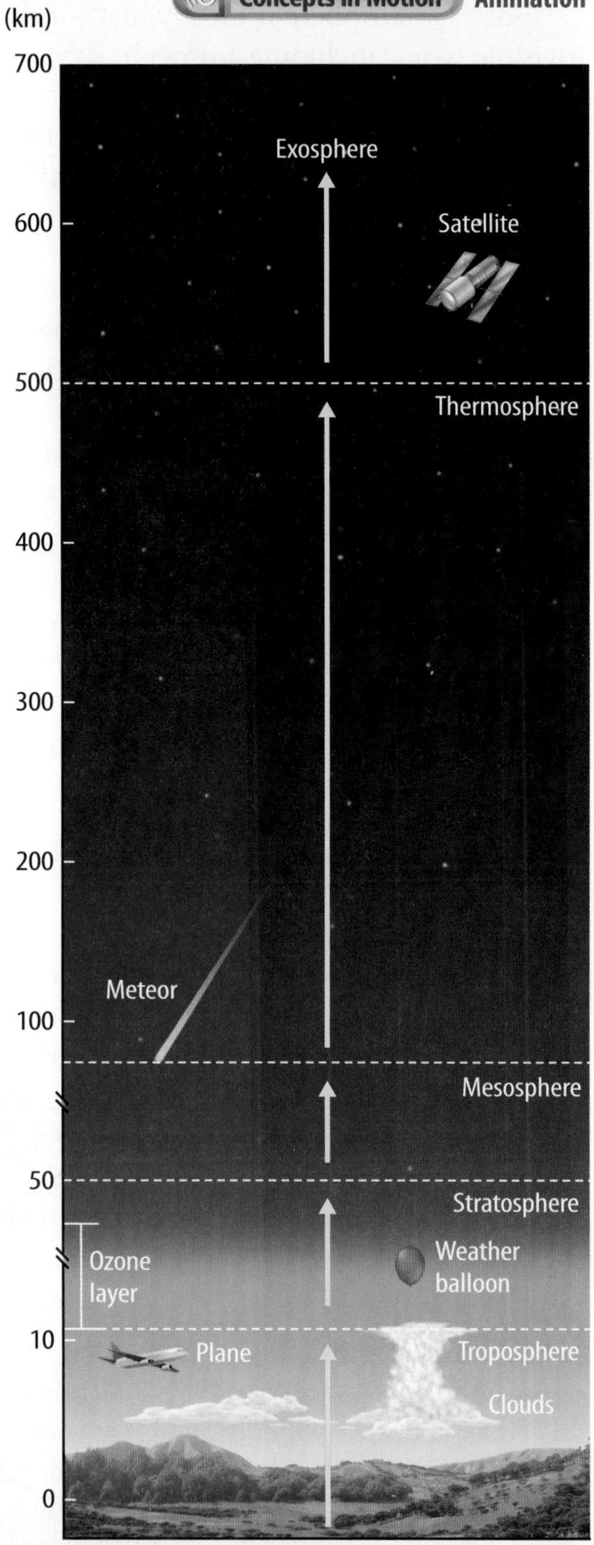

Figure 3 Scientists divide Earth's atmosphere into different layers.

Visual Check In which layer of the atmosphere do planes fly?

Layers of the Atmosphere

The atmosphere has several different layers, as shown in **Figure 3.** Each layer has unique properties, including the composition of gases and how temperature changes with altitude. Notice that the scale between 0–100 km in **Figure 3** is not the same as the scale from 100–700 km. This is so all the layers can be shown in one image.

Troposphere

The atmospheric layer closest to Earth's surface is called the **troposphere** (TRO puh sfihr). Most people spend their entire lives within the troposphere. It extends from Earth's surface to altitudes between 8–15 km. Its name comes from the Greek word *tropos,* which means "change." The temperature in the troposphere decreases as you move away from Earth. The warmest part of the troposphere is near Earth's surface. This is because most sunlight passes through the atmosphere and warms Earth's surface. The warmth is radiated to the troposphere, causing weather.

Reading Check Describe the troposphere.

Stratosphere

The atmospheric layer directly above the troposphere is the **stratosphere** (STRA tuh sfihr). The stratosphere extends from about 15 km to about 50 km above Earth's surface. The lower half of the stratosphere contains the greatest amount of ozone gas. *The area of the stratosphere with a high concentration of ozone* is referred to as the **ozone layer.** The presence of the ozone layer causes increasing stratospheric temperatures with increasing altitude.

An ozone (O_3) molecule differs from an oxygen (O_2) molecule. Ozone has three oxygen atoms instead of two. This difference is important because ozone absorbs the Sun's ultraviolet rays more effectively than oxygen does. Ozone protects Earth from ultraviolet rays that can kill plants, animals, and other organisms and cause skin cancer in humans.

Mesosphere and Thermosphere

As shown in **Figure 3**, the mesosphere extends from the stratosphere to about 85 km above Earth. The thermosphere can extend from the mesopshere to more than 500 km above Earth. Combined, these layers are much broader than the troposphere and the stratosphere, yet only 1 percent of the atmosphere's gas molecules are found in the mesosphere and the thermosphere. Most meteors burn up in these layers instead of striking Earth.

Ionosphere The **ionosphere** *is a region within the mesosphere and thermosphere that contains ions.* Between 60 km and 500 km above Earth's surface, the ionosphere's ions reflect AM radio waves transmitted at ground level. After sunset when ions recombine, this reflection increases. **Figure 4** shows how AM radio waves can travel long distances, especially at night, by bouncing off Earth and the ionosphere.

FOLDABLES
Make a vertical four-tab book using the titles shown. Use it to record similarities and differences among these four layers of the atmosphere. Fold the top half over the bottom and label the outside *Layers of the Atmosphere.*

Radio Waves and the Ionosphere

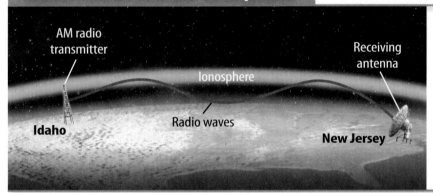

Figure 4 Radio waves can travel long distances in the atmosphere.

Auroras The ionosphere is where stunning displays of colored lights called auroras occur, as shown in **Figure 5**. Auroras are most frequent in the spring and fall, but are best seen when the winter skies are dark. Auroras occur when ions from the Sun strike air molecules, causing them to emit vivid colors of light. People who live in the higher latitudes, nearer to the North Pole and the South Pole, are most likely to see auroras.

Exosphere

The exosphere is the atmospheric layer farthest from Earth's surface. Here, pressure and density are so low that individual gas molecules rarely strike one another. The molecules move at incredibly fast speeds after absorbing the Sun's radiation. The atmosphere does not have a definite edge, and molecules that are part of it can escape the pull of gravity and travel into space.

 Key Concept Check What are the layers of the atmosphere?

▲ **Figure 5** Auroras occur in the ionosphere.

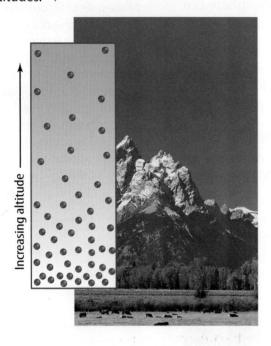

Figure 6 Molecules in the air are closer together near Earth's surface than they are at higher altitudes. ▼

Air Pressure and Altitude

Gravity is the force that pulls all objects toward Earth. When you stand on a scale, you can read your weight. This is because gravity is pulling you toward Earth. Gravity also pulls the atmosphere toward Earth. The pressure that a column of air exerts on anything below it is called air pressure. Gravity's pull on air increases its density. At higher altitudes, the air is less dense. **Figure 6** shows that air pressure is greatest near Earth's surface because the air molecules are closer together. This dense air exerts more force than the less dense air near the top of the atmosphere. Mountain climbers sometimes carry oxygen tanks at high altitudes because fewer oxygen molecules are in the air at high altitudes.

 Reading Check How does air pressure change as altitude increases?

Temperature and Altitude

Figure 7 shows how temperature changes with altitude in the different layers of the atmosphere. If you have ever been hiking in the mountains, you have experienced the temperature cooling as you hike to higher elevations. In the troposphere, temperature decreases as altitude increases. Notice that the opposite effect occurs in the stratosphere. As altitude increases, temperature increases. This is because of the high concentration of ozone in the stratosphere. Ozone absorbs energy from sunlight, which increases the temperature in the stratosphere.

In the mesosphere, as altitude increases, temperature again decreases. In the thermosphere and exosphere, temperatures increase as altitude increases. These layers receive large amounts of energy from the Sun. This energy is spread across a small number of particles, creating high temperatures.

 Key Concept Check How does temperature change as altitude increases?

Figure 7 Temperature differences occur within the layers of the atmosphere. ▼

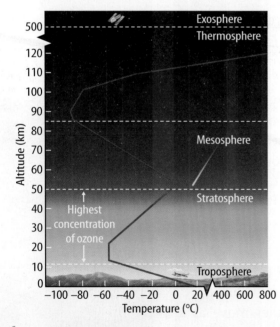

Visual Check Which temperature pattern is most like the troposphere's?

Lesson 1 Review

Assessment — Online Quiz
Inquiry — Virtual Lab

Visual Summary

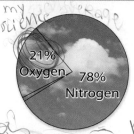

Earth's atmosphere consists of gases that make life possible.

Layers of the atmosphere include the troposphere, the stratosphere, the mesosphere, the thermosphere, and the exosphere.

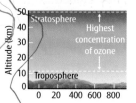

The ozone layer is the area in the stratosphere with a high concentration of ozone.

FOLDABLES

Use your lesson Foldable to review the lesson. Save your Foldable for the project at the end of the chapter.

What do you think NOW?

You first read the statements below at the beginning of the chapter.

1. Air is empty space.
2. Earth's atmosphere is important to living organisms.

Did you change your mind about whether you agree or disagree with the statements? Rewrite any false statements to make them true.

Use Vocabulary

1. The _____ is a thin layer of gases surrounding Earth.

2. The area of the stratosphere that helps protect Earth's surface from harmful ultraviolet rays is the _____.

3. **Define** Using your own words, define *water vapor*.

Understand Key Concepts

4. Which atmospheric layer is closest to Earth's surface?
   A. mesosphere   C. thermosphere
   B. stratosphere   D. troposphere

5. **Identify** the two atmospheric layers in which temperature decreases as altitude increases.

Interpret Graphics

6. **Contrast** Copy and fill in the graphic organizer below to contrast the composition of gases in Earth's early atmosphere and its present-day atmosphere.

Atmosphere	Gases
Early	
Present-day	

7. **Determine** the relationship between air pressure and the water in the glass in the photo below.

Critical Thinking

8. **Explain** three ways the atmosphere is important to living things.

Lesson 1
EVALUATE
415

CAREERS in SCIENCE

A Crack in Earth's Shield

AMERICAN MUSEUM OF NATURAL HISTORY

Scientists discover an enormous hole in the ozone layer that protects Earth.

The ozone layer is like sunscreen, protecting Earth from the Sun's ultraviolet rays. But not all of Earth is covered. Every spring since 1985, scientists have been monitoring a growing hole in the ozone layer above Antarctica.

This surprising discovery was the outcome of years of research from Earth and space. The first measurements of polar ozone levels began in the 1950s, when a team of British scientists began launching weather balloons in Antarctica. In the 1970s, NASA started using satellites to measure the ozone layer from space. Then, in 1985 a close examination of the British team's records indicated a large drop in ozone levels during the Antarctic spring. The levels were so low that the scientists checked and rechecked their instruments before they reported their findings. NASA scientists quickly confirmed the discovery—an enormous hole in the ozone layer over the entire continent of Antarctica. They reported that the hole might have originated as far back as 1976.

Human-made compounds found mostly in chemicals called chlorofluorocarbons, or CFCs, are destroying the ozone layer. During cold winters, molecules released from these compounds are transformed into new compounds by chemical reactions on ice crystals that form in the ozone layer over Antarctica. In the spring, warming by the Sun breaks down the new compounds and releases chlorine and bromine. These chemicals break apart ozone molecules, slowly destroying the ozone layer.

In 1987, CFCs were banned in many countries around the world. Since then, the loss of ozone has slowed and possibly reversed, but a full recovery will take a long time. One reason is that CFCs stay in the atmosphere for more than 40 years. Still, scientists predict the hole in the ozone layer will eventually mend.

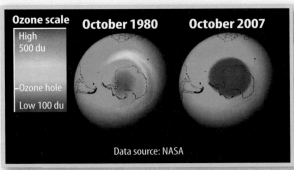

▲ A hole in the ozone layer has developed over Antarctica. Even though it has gotten worse over the years, the hole has not grown as fast as scientists initially thought it would.

Global Warming and the Ozone

Drew Shindell is a NASA scientist investigating the connection between the ozone layer in the stratosphere and the buildup of greenhouse gases throughout the atmosphere. Surprisingly, while these gases warm the troposphere, they are actually causing temperatures in the stratosphere to become cooler. As the stratosphere cools above Antarctica, more clouds with ice crystals form—a key step in the process of ozone destruction. While the buildup of greenhouse gases in the atmosphere may slow the recovery, Shindell still thinks that eventually the ozone layer will heal itself.

It's Your Turn

NEWSCAST Work with a partner to develop three questions about the ozone layer. Research to find the answers. Take the roles of reporter and scientist. Present your findings to the class in a newscast format.

Lesson 2

Energy Transfer in the Atmosphere

Reading Guide

Key Concepts
ESSENTIAL QUESTIONS

- How does energy transfer from the Sun to Earth and the atmosphere?
- How are air circulation patterns within the atmosphere created?

Vocabulary

radiation p. 418
conduction p. 421
convection p. 421
stability p. 422
temperature inversion p. 423

g Multilingual eGlossary

Inquiry What's really there?

Mirages are created as light passes through layers of air that have different temperatures. How does energy create the reflections? What other effects does energy have on the atmosphere?

Inquiry Launch Lab

15 minutes

What happens to air as it warms?

Light energy from the Sun is converted to thermal energy on Earth. Thermal energy powers the weather systems that impact your everyday life.

1. Read and complete a lab safety form.
2. Turn on a **lamp** with an incandescent lightbulb.
3. Place your hands under the light near the lightbulb. What do you feel?
4. Dust your hands with **powder.**
5. Place your hands below the lightbulb and clap them together once.
6. Observe what happens to the particles.

Think About This

1. How might the energy in step 3 move from the lightbulb to your hand?
2. How did the particles move when you clapped your hands?
3. **Key Concept** How did particle motion show you how the air was moving?

Energy from the Sun

The Sun's energy travels 148 million km to Earth in only 8 minutes. How does the Sun's energy get to Earth? It reaches Earth through the process of radiation. **Radiation** *is the transfer of energy by electromagnetic waves.* Ninety-nine percent of the radiant energy from the Sun consists of visible light, ultraviolet light, and infrared radiation.

Visible Light

The majority of sunlight is visible light. Recall that visible light is light that you can see. The atmosphere is like a window to visible light, allowing it to pass through. At Earth's surface it is converted to thermal energy, commonly called heat.

Near-Visible Wavelengths

The wavelengths of ultraviolet (UV) light and infrared radiation (IR) are just beyond the range of visibility to human eyes. UV light has short wavelengths and can break chemical bonds. Excess exposure to UV light will burn human skin and can cause skin cancer. Infrared radiation (IR) has longer wavelengths than visible light. You can sense IR as thermal energy or warmth. Earth absorbs energy from the Sun and then radiates it into the atmosphere as IR.

Reading Check Contrast visible light and ultraviolet light.

Academic Vocabulary

process
(noun) an ordered series of actions

Energy on Earth

As the Sun's energy passes through the atmosphere, some of it is absorbed by gases and particles, and some of it is reflected back into space. As a result, not all the energy coming from the Sun reaches Earth's surface.

Absorption

Study **Figure 8.** Gases and particles in the atmosphere absorb about 20 percent of incoming solar radiation. Oxygen, ozone, and water vapor all absorb incoming ultraviolet light. Water and carbon dioxide in the troposphere absorb some infrared radiation from the Sun. Earth's atmosphere does not absorb visible light. Visible light must be converted to infrared radiation before it can be absorbed.

Reflection

Bright surfaces, especially clouds, **reflect** incoming radiation. Study **Figure 8** again. Clouds and other small particles in the air reflect about 25 percent of the Sun's radiation. Some radiation travels to Earth's surface and is then reflected by land and sea surfaces. Snow-covered, icy, or rocky surfaces are especially reflective. As shown in **Figure 8,** this accounts for about 5 percent of incoming radiation. In all, about 30 percent of incoming radiation is reflected into space. This means that, along with the 20 percent of incoming radiation that is absorbed in the atmosphere, Earth's surface only receives and absorbs about 50 percent of incoming solar radiation.

> **SCIENCE USE V. COMMON USE**
> **reflect**
> *Science Use* to return light, heat, sound, and so on, after it strikes a surface
> *Common Use* to think quietly and calmly

Figure 8 Some of the energy from the Sun is reflected or absorbed as it passes through the atmosphere.

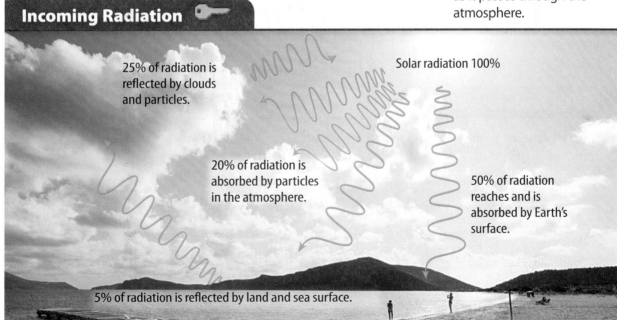

Incoming Radiation

- 25% of radiation is reflected by clouds and particles.
- Solar radiation 100%
- 20% of radiation is absorbed by particles in the atmosphere.
- 50% of radiation reaches and is absorbed by Earth's surface.
- 5% of radiation is reflected by land and sea surface.

Visual Check What percent of incoming radiation is absorbed by gases and particles in the atmosphere?

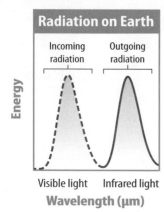

▲ **Figure 9** The amount of solar energy absorbed by Earth and its atmosphere is equal to the amount of energy Earth radiates back into space.

Radiation Balance

The Sun's radiation heats Earth. So, why doesn't Earth get hotter and hotter as it continues to receive radiation from the Sun? There is a balance between the amount of incoming radiation from the Sun and the amount of outgoing radiation from Earth.

The land, water, plants, and other organisms absorb solar radiation that reaches Earth's surface. The radiation absorbed by Earth is then re-radiated, or bounced back, into the atmosphere. Most of the energy radiated from Earth is infrared radiation, which heats the atmosphere. **Figure 9** shows that the amount of radiation Earth receives from the Sun is the same as the amount Earth radiates into the outer atmosphere. Earth absorbs the Sun's energy and then radiates that energy away until a balance is achieved.

The Greenhouse Effect

As shown in **Figure 10,** the glass of a greenhouse allows light to pass through, where it is converted to infrared energy. The glass prevents the IR from escaping and it warms the greenhouse. Some of the gases in the atmosphere, called greenhouse gases, act like the glass of a greenhouse. They allow sunlight to pass through, but they prevent some of Earth's IR energy from escaping. Greenhouse gases in Earth's atmosphere trap IR and direct it back to Earth's surface. This causes an additional buildup of thermal energy at Earth's surface. The gases that trap IR best are water vapor (H_2O), carbon dioxide (CO_2), and methane (CH_4).

 Reading Check Describe the greenhouse effect.

The Greenhouse Effect

Figure 10 Some of the outgoing radiation is directed back toward Earth's surface by greenhouse gases.

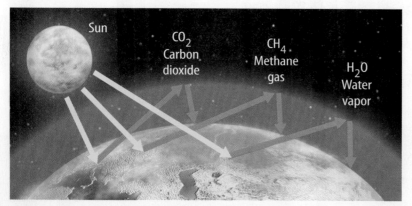

Chapter 12
EXPLAIN

Thermal Energy Transfer

Recall that there are three types of thermal energy transfer—radiation, conduction, and convection. All three occur in the atmosphere. Recall that radiation is the process that transfers energy from the Sun to Earth.

Conduction

Thermal energy always moves from an object with a higher temperature to an object with a lower temperature. **Conduction** *is the transfer of thermal energy by collisions between particles of matter.* Particles must be close enough to touch to transfer energy by conduction. Touching the pot of water, shown in **Figure 11,** would transfer energy from the pot to your hand. Conduction occurs where the atmosphere touches Earth.

Convection

As molecules of air close to Earth's surface are heated by conduction, they spread apart, and air becomes less dense. Less dense air rises, transferring thermal energy to higher altitudes. *The transfer of thermal energy by the movement of particles within matter is called* **convection.** Convection can be seen in **Figure 11** as the boiling water circulates and steam rises.

Latent Heat

More than 70 percent of Earth's surface is covered by a highly unique substance—water! Water is the only substance that can exist as a solid, a liquid, and a gas within Earth's temperature ranges. Recall that latent heat is exchanged when water changes from one phase to another, as shown in **Figure 12.** Latent heat energy is transferred from Earth's surface to the atmosphere.

 Key Concept Check How does energy transfer from the Sun to Earth and the atmosphere?

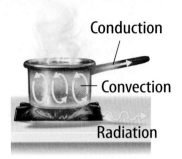

▲ **Figure 11** Energy is transferred through conduction, convection, and radiation.

Review
Personal Tutor

WORD ORIGIN
conduction
from Latin *conducere*, means "to bring together"

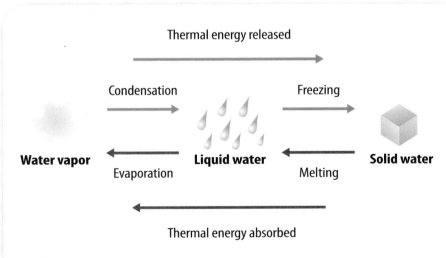

Figure 12 Water releases or absorbs thermal energy during phase changes.

FOLDABLES

Fold a sheet of paper to make a four-column, four-row table and label as shown. Use it to record information about thermal energy transfer.

Circulating Air

Figure 13 Rising warm air is replaced by cooler, denser air that sinks beside it.

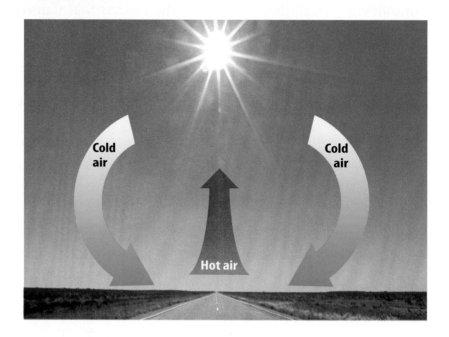

Figure 14 Lens-shaped lenticular clouds form when air rises with a mountain wave. ▼

Mountain Wave

Circulating Air

You've read that energy is transferred through the atmosphere by convection. On a hot day, air that is heated becomes less dense. This creates a pressure difference. Cool, denser air pushes the warm air out of the way. The warm air is replaced by the more dense air, as shown in **Figure 13**. The warm air is often pushed upward. Warmer, rising air is always accompanied by cooler, sinking air.

Air is constantly moving. For example, wind flowing into a mountain range rises and flows over it. After reaching the top, the air sinks. This up-and-down motion sets up an atmospheric phenomenon called a mountain wave. The upward moving air within mountain waves creates lenticular (len TIH kyuh lur) clouds, shown in **Figure 14**. Circulating air affects weather and climate around the world.

Key Concept Check How are air circulation patterns within the atmosphere created?

Stability

When you stand in the wind, your body forces some of the air to move above you. The same is true for hills and buildings. Conduction and convection also cause air to move upward. **Stability** *describes whether circulating air motions will be strong or weak*. When air is unstable, circulating motions are strong. During stable conditions, circulating motions are weak.

Normal conditions

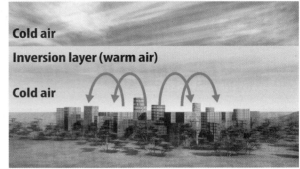

Temperature inversion

Unstable Air and Thunderstorms Unstable conditions often occur on warm, sunny afternoons. During unstable conditions, ground-level air is much warmer than higher-altitude air. As warm air rises rapidly in the atmosphere, it cools and forms large, tall clouds. Latent heat, released as water vapor changes from a gas to a liquid, adds to the instability, and produces a thunderstorm.

Reading Check Relate unstable air to the formation of thunderstorms.

Stable Air and Temperature Inversions Sometimes ground-level air is nearly the same temperature as higher-altitude air. During these conditions, the air is stable, and circulating motions are weak. A temperature inversion can occur under these conditions. *A* **temperature inversion** *occurs in the troposphere when temperature increases as altitude increases.* During a temperature inversion, a layer of cooler air is trapped by a layer of warmer air above it, as shown in **Figure 15.** Temperature inversions prevent air from mixing and can trap pollution in the air close to Earth's surface.

Figure 15 A temperature inversion occurs when cooler air is trapped beneath warmer air.

Visual Check How do conditions during a temperature inversion differ from normal conditions?

Inquiry MiniLab

20 minutes

Can you identify a temperature inversion?

You've read that a temperature inversion is a reversal of normal temperature conditions in the troposphere. What do data from a temperature inversion look like on a graph?

Analyze and Conclude

1. **Describe** the information presented in the graph. How do the graph's lines differ?

2. **Analyze** Which graph line represents normal conditions in the troposphere? Which represents a temperature inversion? Explain your answers in your Science Journal.

3. **Key Concept** From the graph, what pattern does a temperature inversion have?

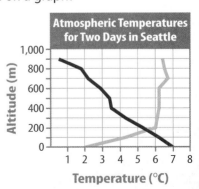

Lesson 2
EXPLAIN

Lesson 2 Review

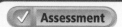 Assessment   Online Quiz

Visual Summary

Not all radiation from the Sun reaches Earth's surface.

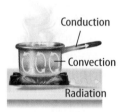

Thermal energy transfer in the atmosphere occurs through radiation, conduction, and convection.

Temperature inversions prevent air from mixing and can trap pollution in the air close to Earth's surface.

FOLDABLES

Use your lesson Foldable to review the lesson. Save your Foldable for the project at the end of the chapter.

What do you think NOW?

You first read the statements below at the beginning of the chapter.

3. All of the energy from the Sun reaches Earth's surface.
4. Earth emits energy back into the atmosphere.

Did you change your mind about whether you agree or disagree with the statements? Rewrite any false statements to make them true.

Use Vocabulary

1. The property of the atmosphere that describes whether circulating air motions will be strong or weak is called _____.

2. **Define** *conduction* in your own words.

3. _____ is the transfer of thermal energy by the movement of particles within matter.

Understand Key Concepts

4. Which statement is true?
   A. The Sun's energy is completely blocked by Earth's atmosphere.
   B. The Sun's energy passes through the atmosphere without warming it significantly.
   C. The Sun's IR energy is absorbed by greenhouse gases.
   D. The Sun's energy is primarily in the UV range.

5. **Distinguish** between conduction and convection.

Interpret Graphics

6. **Explain** how greenhouses gases affect temperatures on Earth.

7. **Sequence** Copy and fill in the graphic organizer below to describe how energy from the Sun is absorbed in Earth's atmosphere.

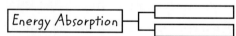

Critical Thinking

8. **Suggest** a way to keep a parked car cool on a sunny day.

9. **Relate** temperature inversions to air stability.

Inquiry Skill Practice: Compare and Contrast

30 minutes

Materials

candle

metal rod

glass rod

wooden dowel

500-mL beaker

ice

bowls (2)

lamp

glass cake pan

food coloring

250-mL beaker

Safety

Can you conduct, convect, and radiate?

After solar radiation reaches Earth, the molecules closest to Earth transfer thermal energy from molecule to molecule by conduction. The newly warmed air becomes less dense and moves through the process of convection.

Learn It

When you **compare and contrast** two or more things, you look for similarities and differences between them. When you **compare** two things, you look for the similarities, or how they are the same. When you **contrast** them, you look for how they are different from each other.

Try It

1. Read and complete a lab safety form.
2. Drip a small amount of melted candle wax onto one end of a metal rod, a glass rod, and a wooden dowel.
3. Place a 500-mL beaker on the lab table. Have your teacher add 350 mL of very hot water. Place the ends of the rods without candle wax in the water. Set aside.

4. Place an ice cube into each of two small bowls labeled A and B.
5. Place bowl A under a lamp with a 60- or 75-watt lightbulb. Place the light source 10 cm above the bowl. Turn on the lamp. Set bowl B aside.
6. Fill a glass cake pan with room-temperature water to a level of 2 cm. Put 2–3 drops of red food coloring into a 250-mL beaker of very hot water. Put 2–3 drops of blue food coloring into a 250-mL beaker of very cold water and ice cubes. Carefully pour the hot water into one end of the pan. Slowly pour the very cold water into the same end of the pan. Observe what happens from the side of the pan. Record your observations in your Science Journal.
7. Observe the candle wax on the rods in the hot water and the ice cubes in the bowls.

Apply It

8. What happened to the candle wax? Identify the type of energy transfer.
9. Which ice cube melted the most in the bowls? Identify the type of energy transfer that melted the ice.
10. Compare and contrast how the hot and cold water behaved in the pan. Identify the type of energy transfer.
11. 🔑 **Key Concept** Explain how each part of the lab models radiation, conduction, or convection.

Lesson 3

Air Currents

Reading Guide

Key Concepts 🔑
ESSENTIAL QUESTIONS

- How does uneven heating of Earth's surface result in air movement?
- How are air currents on Earth affected by Earth's spin?
- What are the main wind belts on Earth?

Vocabulary
wind p. 427
trade winds p. 429
westerlies p. 429
polar easterlies p. 429
jet stream p. 429
sea breeze p. 430
land breeze p. 430

 Multilingual eGlossary

 Video

What's Science Got to do With It?

Inquiry How does air push these blades?

If you have ever ridden a bicycle into a strong wind, you know the movement of air can be a powerful force. Some areas of the world have more wind than others. What causes these differences? What makes wind?

Launch Lab

15 minutes

Why does air move?

Early sailors relied on wind to move their ships around the world. Today, wind is used as a renewable source of energy. In the following activity, you will explore what causes air to move.

1. Read and complete a lab safety form.
2. Inflate a **balloon.** Do not tie it. Hold the neck of the balloon closed.
3. Describe how the inflated balloon feels.
4. Open the neck of the balloon without letting go of the balloon. Record your observations of what happens in your Science Journal.

Think About This

1. What caused the inflated balloon surface to feel the way it did when the neck was closed?
2. What caused the air to leave the balloon when the neck was opened?
3. **Key Concept** Why didn't outside air move into the balloon when the neck was opened?

Global Winds

There are great wind belts that circle the globe. The energy that causes this massive movement of air originates at the Sun. However, wind patterns can be global or local.

Unequal Heating of Earth's Surface

The Sun's energy warms Earth. However, the same amount of energy does not reach all of Earth's surface. The amount of energy an area gets depends largely on the Sun's angle. For example, energy from the rising or setting Sun is not very intense. But Earth heats up quickly when the Sun is high in the sky.

In latitudes near the equator—an area referred to as the tropics—sunlight strikes Earth's surface at a high angle—nearly 90°—year round. As a result, in the tropics there is more sunlight per unit of surface area. This means that the land, the water, and the air at the equator are always warm.

At latitudes near the North Pole and the South Pole, sunlight strikes Earth's surface at a low angle. Sunlight is now spread over a larger surface area than in the tropics. As a result, the poles receive very little energy per unit of surface area and are cooler.

Recall that differences in density cause warm air to rise. Warm air puts less pressure on Earth than cooler air. Because it's so warm in the tropics, air pressure is usually low. Over colder areas, such as the North Pole and the South Pole, air pressure is usually high. This difference in pressure creates wind. **Wind** *is the movement of air from areas of high pressure to areas of low pressure.* Global wind belts influence both climate and weather on Earth.

 Key Concept Check How does uneven heating of Earth's surface result in air movement?

Global Wind Belts

Figure 16 Three cells in each hemisphere move air through the atmosphere.

Visual Check Which wind belt do you live in?

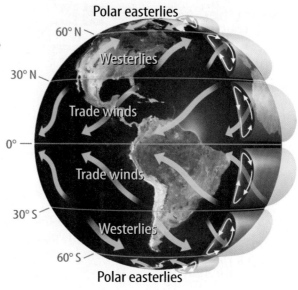

Global Wind Belts

Figure 16 shows the three-cell model of circulation in Earth's atmosphere. In the northern hemisphere, hot air in the cell nearest the equator moves to the top of the troposphere. There, the air moves northward until it cools and moves back to Earth's surface near 30° latitude. Most of the air in this convection cell then returns to Earth's surface near the equator.

The cell at the highest northern latitudes is also a convection cell. Air from the North Pole moves toward the equator along Earth's surface. The cooler air pushes up the warmer air near 60° latitude. The warmer air then moves northward and repeats the cycle. The cell between 30° and 60° latitude is not a convection cell. Its motion is driven by the other two cells, in a motion similar to a pencil that you roll between your hands. Three similar cells exist in the southern hemisphere. These cells help generate the global wind belts.

The Coriolis Effect

What happens when you throw a ball to someone across from you on a moving merry-go-round? The ball appears to curve because the person catching the ball has moved. Similarly, Earth's rotation causes moving air and water to appear to move to the right in the northern hemisphere and to the left in the southern hemisphere. This is called the Coriolis effect. The contrast between high and low pressure and the Coriolis effect creates distinct wind patterns, called prevailing winds.

Key Concept Check How are air currents on Earth affected by Earth's spin?

FOLDABLES

Make a shutterfold. As illustrated, draw Earth and the three cells found in each hemisphere on the inside of the shutterfold. Describe each cell and explain the circulation of Earth's atmosphere. On the outside, label the global wind belts.

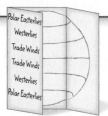

Prevailing Winds

The three global cells in each hemisphere create northerly and southerly winds. When the Coriolis effect acts on the winds, they blow to the east or the west, creating relatively steady, predictable winds. Locate the trade winds in **Figure 16.** The **trade winds** are steady winds that flow from east to west between 30°N latitude and 30°S latitude.

At about 30°N and 30°S air cools and sinks. This creates areas of high pressure and light, calm winds, called the doldrums. Sailboats without engines can be stranded in the doldrums.

The prevailing **westerlies** are steady winds that flow from west to east between latitudes 30°N and 60°N, and 30°S and 60°S. This region is also shown in **Figure 15.** The **polar easterlies** are cold winds that blow from the east to the west near the North Pole and the South Pole.

 Key Concept Check What are the main wind belts on Earth?

Jet Streams

Near the top of the troposphere is a narrow band of high winds called the **jet stream**. Shown in **Figure 17,** jet streams flow around Earth from west to east, often making large loops to the north or the south. Jet streams influence weather as they move cold air from the poles toward the tropics and warm air from the tropics toward the poles. Jet streams can move at speeds up to 300 km/h and are more unpredictable than prevailing winds.

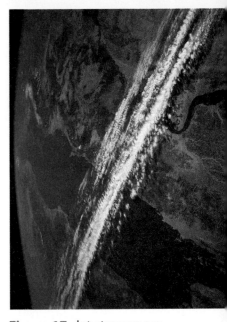

Figure 17 Jet streams are thin bands of high wind speed. The clouds seen here have condensed within a cooler jet stream.

Inquiry MiniLab
20 minutes

Can you model the Coriolis effect?

Earth's rotation causes the Coriolis effect. It affects the movement of water and air on Earth.

1. Read and complete a lab safety form.
2. Draw dot A in the center of a piece of **foamboard.** Draw dot B along the outer edge of the foamboard.
3. Roll a **table-tennis ball** from dot A to dot B. Record your observations in your Science Journal.
4. Center the foamboard on a **turntable**. Have your partner rotate the foamboard at a medium speed. Roll the ball along the same path. Record your observations.

Analyze and Conclude

1. **Contrast** the path of the ball when the foamboard was not moving to when it was spinning.
2.  **Key Concept** How might air moving from the North Pole to the equator travel due to Earth's rotation?

Lesson 3
EXPLAIN

Local Winds

You have just read that global winds occur because of pressure differences around the globe. In the same way, local winds occur whenever air pressure is different from one location to another.

Sea and Land Breezes

Anyone who has spent time near a lake or an ocean shore has probably experienced the connection between temperature, air pressure, and wind. A **sea breeze** *is wind that blows from the sea to the land due to local temperature and pressure differences.* **Figure 18** shows how sea breezes form. On sunny days, land warms up faster than water does. The air over the land warms by conduction and rises, creating an area of low pressure. The air over the water sinks, creating an area of high pressure because it is cooler. The differences in pressure over the warm land and the cooler water result in a cool wind that blows from the sea onto land.

A **land breeze** *is a wind that blows from the land to the sea due to local temperature and pressure differences.* **Figure 18** shows how land breezes form. At night, the land cools more quickly than the water. Therefore, the air above the land cools more quickly than the air over the water. As a result, an area of lower pressure forms over the warmer water. A land breeze then blows from the land toward the water.

 Reading Check Compare and contrast sea breezes and land breezes.

Figure 18 Sea breezes and land breezes are created as part of a large reversible convection current.

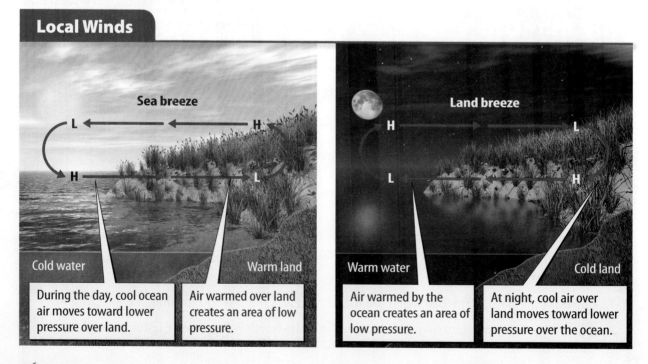

Local Winds

Sea breeze

Cold water — During the day, cool ocean air moves toward lower pressure over land.

Warm land — Air warmed over land creates an area of low pressure.

Land breeze

Warm water — Air warmed by the ocean creates an area of low pressure.

Cold land — At night, cool air over land moves toward lower pressure over the ocean.

Visual Check Sequence the steps involved in the formation of a land breeze.

Lesson 3 Review

Assessment | Online Quiz

Visual Summary

Wind is created by pressure differences between one location and another.

Prevailing winds in the global wind belts are the trade winds, the westerlies, and the polar easterlies.

Sea breezes and land breezes are examples of local winds.

FOLDABLES

Use your lesson Foldable to review the lesson. Save your Foldable for the project at the end of the chapter.

What do you think NOW?

You first read the statements below at the beginning of the chapter.

5. Uneven heating in different parts of the atmosphere creates air circulation patterns.

6. Warm air sinks and cold air rises.

Did you change your mind about whether you agree or disagree with the statements? Rewrite any false statements to make them true.

Use Vocabulary

1. The movement of air from areas of high pressure to areas of low pressure is _____.

2. A(n) _____ is wind that blows from the sea to the land due to local temperature and pressure differences.

3. **Distinguish** between westerlies and trade winds.

Understand Key Concepts

4. Which does NOT affect global wind belts?
    A. air pressure
    B. land breezes
    C. the Coriolis effect
    D. the Sun

5. **Relate** Earth's spinning motion to the Coriolis effect.

Interpret Graphics

Use the image below to answer question 6.

6. **Explain** a land breeze.

7. **Organize** Copy and fill in the graphic organizer below to summarize Earth's global wind belts.

Wind Belt	Description
Trade winds	
Westerlies	
Polar easterlies	

Critical Thinking

8. **Infer** what would happen without the Coriolis effect.

9. **Explain** why the wind direction is often the same in Hawaii as it is in Greenland.

Lesson 3 • 431
EVALUATE

Inquiry Skill Practice: Model

30 minutes

Can you model global wind patterns?

In each hemisphere, air circulates in specific patterns. Recall that scientists use the three-cell model to describe these circulation cells. General circulation of the atmosphere produces belts of prevailing winds around the world. In this activity, you will make a **model** of the main circulation cells in Earth's atmosphere.

Materials

ribbons

globe

permanent marker

scissors

transparent tape

Safety

Learn It

Making a **model** can help you visualize how a process works. Scientists use models to represent processes that may be difficult to see in real time. Sometimes a model represents something too small to see with the unaided eye, such as a model of an atom. Other models, such as one of the solar system, represent something that is too large to see from one location.

Try It

1. Read and complete a lab safety form.

2. Refer to **Figure 16** to make your model.

3. Choose one color of ribbon for the circulation cells. Make a separate loop of ribbon long enough to cover the latitude boundaries of each cell. Draw arrows on each ribbon to show the direction that the air flows in that cell. Make one loop for each cell in the northern hemisphere and one for each in the southern hemisphere. Tape your "cells" onto the globe.

4. Choose different-colored ribbons to model each of these wind belts: trade winds, westerlies, and polar easterlies, in both hemispheres. Draw arrows on each ribbon to show the direction that the wind blows. Tape the ribbons on the globe.

5. Create a color key to identify each cell and its corresponding wind type.

Apply It

6. Explain how your model represents the three-cell model used by scientists. How does your model differ from actual air movement in the atmosphere?

7. Explain why you cannot accurately model the global winds with this model.

8. **Key Concept** Explain how latitude affects global winds.

Lesson 4

Air Quality

Reading Guide

Key Concepts
ESSENTIAL QUESTIONS

- How do humans impact air quality?
- Why do humans monitor air quality standards?

Vocabulary
air pollution p. 434
acid precipitation p. 435
photochemical smog p. 435
particulate matter p. 436

 Multilingual eGlossary

 Video BrainPOP®

Inquiry How did this happen?

Air pollution can be trapped near Earth's surface during a temperature inversion. This is especially common in cities located in valleys and surrounded by mountains. What do you think the quality of the air is like on a day like this one? Where does pollution come from?

Lesson 4
ENGAGE
433

Inquiry Launch Lab

20 minutes

How does acid rain form?

Vehicles, factories, and power plants release chemicals into the atmosphere. When these chemicals combine with water vapor, they can form acid rain.

1. Read and complete a lab safety form.
2. Half-fill a **plastic cup** with **distilled water**.
3. Dip a strip of **pH paper** into the water. Use a **pH color chart** to determine the pH of the distilled water. Record the pH in your Science Journal.
4. Use a **dropper** to add **lemon juice** to the water until the pH equals that of acid rain. Swirl and test the pH each time you add 5 drops of the lemon juice to the mixture.

Substances	pH
Hydrochloric acid	0.0
Lemon juice	2.3
Vinegar	2.9
Tomato juice	4.1
Coffee (black)	5.0
Acid rain	5.6
Rainwater	6.5
Milk	6.6
Distilled water	7.0
Blood	7.4
Baking soda solution	8.4
Toothpaste	9.9
Household ammonia	11.9
Sodium hydroxide	14.0

Think About This

1. A strong acid has a pH between 0 and 2. How does the pH of lemon juice compare to the pH of other substances? Is acid rain a strong acid?
2. **Key Concept** Why might scientists monitor the pH of rain?

Sources of Air Pollution

The contamination of air by harmful substances including gases and smoke is called **air pollution**. Air pollution is harmful to humans and other living things. Years of exposure to polluted air can weaken a human's immune system. Respiratory diseases such as asthma can be caused by air pollution.

Air pollution comes from many sources. Point-source pollution is pollution that comes from an identifiable source. Examples of point sources include smokestacks of large factories, such as the one shown in **Figure 19,** and electric power plants that burn fossil fuels. They release tons of polluting gases and particles into the air each day. An example of natural point-source pollution is an erupting volcano.

Nonpoint-source pollution is pollution that comes from a widespread area. One example of pollution from a nonpoint-source is air pollution in a large city. This is considered nonpoint-source pollution because it cannot be traced back to one source. Some bacteria found in swamps and marshes are examples of natural sources of nonpoint-source pollution.

Key Concept Check Compare point-source and nonpoint-source pollution.

Figure 19 One example of point-source pollution is a factory smoke stack.

Causes and Effects of Air Pollution

The harmful effects of air pollution are not limited to human health. Some pollutants, including ground-level ozone, can damage plants. Air pollution can also cause serious damage to human-made structures. Sulfur dioxide pollution can discolor stone, corrode metal, and damage paint on cars.

Acid Precipitation

When sulfur dioxide and nitrogen oxides combine with moisture in the atmosphere and form precipitation that has a pH lower than that of normal rainwater, it is called **acid precipitation.** Acid precipitation includes acid rain, snow, and fog. It affects the chemistry of water in lakes and rivers. This can harm the organisms living in the water. Acid precipitation damages buildings and other structures made of stone. Natural sources of sulfur dioxide include volcanoes and marshes. However, the most common sources of sulfur dioxide and nitrogen oxides are automobile exhausts and factory and power plant smoke.

Smog

Photochemical smog *is air pollution that forms from the interaction between chemicals in the air and sunlight.* Smog forms when nitrogen dioxide, released in gasoline engine exhaust, reacts with sunlight. A series of chemical reactions produces ozone and other compounds that form smog. Recall that ozone in the stratosphere helps protect organisms from the Sun's harmful rays. However, ground-level ozone can damage the tissues of plants and animals. Ground-level ozone is the main component of smog. Smog in urban areas reduces visibility and makes air difficult to breathe. **Figure 20** shows New York City on a clear day and on a smoggy day.

 Key Concept Check How do humans impact air quality?

Make a horizontal three-tab Foldable and label it as shown. Use it to organize your notes about the formation of air pollution and its effects. Fold the right and left thirds over the center and label the outside *Types of Air Pollution.*

Figure 20 Smog can be observed as haze or a brown tint in the atmosphere.

Particulate Pollution

Although you can't see them, over 10,000 solid or liquid particles are in every cubic centimeter of air. A cubic centimeter is about the size of a sugar cube. This type of pollution is called particulate matter. **Particulate** (par TIH kyuh lut) **matter** *is a mixture of dust, acids, and other chemicals that can be hazardous to human health.* The smallest particles are the most harmful. These particles can be inhaled and can enter your lungs. They can cause asthma, bronchitis, and lead to heart attacks. Children and older adults are most likely to experience health problems due to particulate matter.

Particulate matter in the atmosphere absorbs and scatters sunlight. This can create haze. Haze particles scatter light, make things blurry, and reduce visibility.

> **WORD ORIGIN**
> particulate
> from Latin *particula*, means "small part"

Movement of Air Pollution

Wind can influence the effects of air pollution. Because air carries pollution with it, some wind patterns cause more pollution problems than others. Weak winds or no wind prevents pollution from mixing with the surrounding air. During weak wind conditions, pollution levels can become dangerous.

For example, the conditions in which temperature inversions form are weak winds, clear skies, and longer winter nights. As land cools at night, the air above it also cools. Calm winds, however, prevent cool air from mixing with warm air above it. **Figure 21** shows how cities located in valleys experience a temperature inversion. Cool air, along with the pollution it contains, is trapped in valleys. More cool air sinks down the sides of the mountain, further preventing layers from mixing. The pollution in the photo at the beginning of the lesson was trapped due to a temperature inversion.

Figure 21 At night, cool air sinks down the mountain sides, trapping pollution in the valley below.

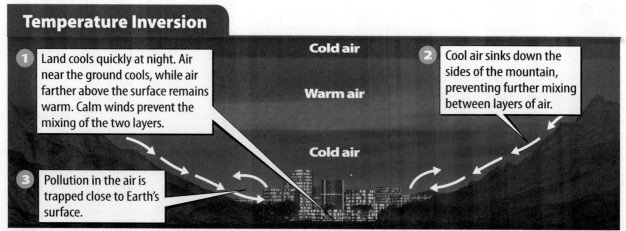

Temperature Inversion

1. Land cools quickly at night. Air near the ground cools, while air farther above the surface remains warm. Calm winds prevent the mixing of the two layers.

2. Cool air sinks down the sides of the mountain, preventing further mixing between layers of air.

3. Pollution in the air is trapped close to Earth's surface.

Visual Check How is pollution trapped by a temperature inversion?

Maintaining Healthful Air Quality

Preserving the quality of Earth's atmosphere requires the cooperation of government officials, scientists, and the public. The Clean Air Act is an example of how government can help fight pollution. Since the Clean Air Act became law in 1970, steps have been taken to reduce automobile emissions. Pollutant levels have decreased significantly in the United States. Despite these advances, serious problems still remain. The amount of ground-level ozone is still too high in many large cities. Also, acid precipitation produced by air pollutants continues to harm organisms in lakes, streams, and forests.

Air Quality Standards

The Clean Air Act gives the U.S. government the power to set air quality standards. The standards protect humans, animals, plants, and buildings from the harmful effects of air pollution. All states are required to make sure that pollutants, such as carbon monoxide, nitrogen oxides, particulate matter, ozone, and sulfur dioxide, do not exceed harmful levels.

 Reading Check What is the Clean Air Act?

Monitoring Air Pollution

Pollution levels are continuously monitored by hundreds of instruments in all major U.S. cities. If the levels are too high, authorities may advise people to limit outdoor activities.

 MiniLab 15 minutes

Can being out in fresh air be harmful to your health?

Are you going to be affected if you play tennis for a couple hours, go biking with your friends, or even just lie on the beach? Even if you have no health problems related to your respiratory system, you still need to be aware of the quality of air in your area of activity for the day.

Analyze and Conclude

1. Which values on the AQI indicate that the air quality is good?
2. At what value is the air quality unhealthful for anyone who may have allergies and respiratory disorders?
3. Which values would be considered as warnings of emergency conditions?
4. **Key Concept** The quality of air in different areas changes throughout the day. Explain how you can use the AQI to help you know when you should limit your outdoor activity.

Air Quality Index (AQI) Values	Levels of Health Concern
0 to 50	Good
51 to 100	Moderate
101 to 150	Unhealthful for Sensitive Groups
151 to 200	Unhealthful
201 to 300	Very Unhealthful
301 to 500	Hazardous

Air Quality Trends

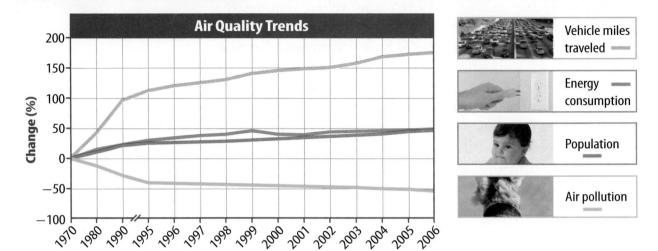

Figure 22 Pollution emissions have declined, even though the population is increasing.

Math Skills

Use Graphs

The graph above shows the percent change in four different pollution factors from 1970 through 2006. All values are based on the 0 percent amount in 1970. For example, from 1970 to 1990, the number of vehicle miles driven increased by 100 percent, or the vehicle miles doubled. Use the graph to infer which factors might be related.

Practice

1. What was the percent change in population between 1970 and 2006?
2. What other factor changed by about the same amount during that period?

Review
- Math Practice
- Personal Tutor

Air Quality Trends

Over the last several decades, air quality in U.S. cities has improved, as shown in **Figure 22.** Even though some pollution-producing processes have increased, such as burning fossil fuels and traveling in automobiles, levels of certain air pollutants have decreased. Airborne levels of lead and carbon monoxide have decreased the most. Levels of sulfur dioxide, nitrogen oxide, and particulate matter have also decreased.

However, ground-level ozone has not decreased much. Why do ground-level ozone trends lag behind those of other pollutants? Recall that ozone can be created from chemical reactions involving automobile exhaust. The increase in the amount of ground-level ozone is because of the increase in the number of miles traveled by vehicles.

 Key Concept Check Why do humans monitor air quality standards?

Indoor Air Pollution

Not all air pollution is outdoors. The air inside homes and other buildings can be as much as 50 times more polluted than outdoor air! The quality of indoor air can impact human health much more than outdoor air quality.

Indoor air pollution comes from many sources. Tobacco smoke, cleaning products, pesticides, and fireplaces are some common sources. Furniture upholstery, carpets, and foam insulation also add pollutants to the air. Another indoor air pollutant is radon, an odorless gas given off by some soil and rocks. Radon leaks through cracks in a building's foundation and sometimes builds up to harmful levels inside homes. Harmful effects of radon come from breathing its particles.

Lesson 4 Review

 Assessment   Online Quiz

Visual Summary

Air pollution comes from point sources, such as factories, and nonpoint sources, such as automobiles.

Photochemical smog contains ozone, which can damage tissues in plants and animals.

FOLDABLES

Use your lesson Foldable to review the lesson. Save your Foldable for the project at the end of the chapter.

What do you think NOW?

You first read the statements below at the beginning of the chapter.

7. If no humans lived on Earth, there would be no air pollution.

8. Pollution levels in the air are not measured or monitored.

Did you change your mind about whether you agree or disagree with the statements? Rewrite any false statements to make them true.

Use Vocabulary

1. **Define** *acid precipitation* in your own words.

2. _____ forms when chemical reactions combine pollution with sunlight.

3. The contamination of air by harmful substances, including gases and smoke, is _____.

Understand Key Concepts

4. Which is NOT true about smog?
   A. It contains nitrogen oxide.
   B. It contains ozone.
   C. It reduces visibility.
   D. It is produced only by cars.

5. **Describe** two ways humans add pollution to the atmosphere.

6. **Assess** whether urban or rural areas are more likely to have high levels of smog.

7. **Identify** and describe the law designed to reduce air pollution.

Interpret Graphics

8. **Compare and Contrast** Copy and fill in the graphic organizer below to compare and contrast details of smog and acid precipitation.

	Similarities	Differences
Smog		
Acid Precipitation		

Critical Thinking

9. **Describe** how conduction and convection are affected by paving over a grass field.

Math Skills
Math Practice

10. Based on the graph on the opposite page, what was the total percent change in air pollution between 1970 and 2006?

Inquiry Lab

40 minutes

Radiant Energy Absorption

Materials

thermometer

sand

500-mL beaker

lamp

stopwatch

paper towels

spoon

potting soil

clay

Safety

Ultimately, the Sun is the source of energy for Earth. Energy from the Sun moves through the atmosphere and is absorbed and reflected from different surfaces on Earth. Light surfaces reflect energy, and dark surfaces absorb energy. Both land and sea surfaces absorb energy from the Sun, and air in contact with these surfaces is warmed through conduction.

Ask a Question
Which surfaces on Earth absorb the most energy from the Sun?

Make Observations

1. Read and complete a lab safety form.
2. Make a data table in your Science Journal to record your observations of energy transfer. Include columns for Type of Surface, Temperature Before Heating, and Temperature After Heating.

3. Half-fill a 500-mL beaker with sand. Place a thermometer in the sand and carefully add enough sand to cover the thermometer bulb—about 2 cm deep. Keep the bulb under the sand for 1 minute. Record the temperature in the data table.
4. Place the beaker under the light source. Record the temperature after 10 minutes.
5. Repeat steps 3 and 4 using soil and water.

Form a Hypothesis

6. Use the data in your table to form a hypothesis stating which surfaces on Earth, such as forests, wheat fields, lakes, snowy mountain tops, and deserts, will absorb the most radiant energy.

Test Your Hypothesis

7. Decide what materials could be used to mimic the surfaces on Earth from your hypothesis.

8. Repeat the experiment with materials approved by the teacher to test your hypothesis.

9. Examine your data. Was your hypothesis supported? Why or why not?

Analyze and Conclude

10. **Infer** which types of areas on Earth absorb the most energy from the Sun.

11. **Think Critically** When areas of Earth are changed so they become more likely to reflect or absorb energy from the Sun, how might these changes affect conduction and convection in the atmosphere?

12. **The Big Idea** Explain how thermal energy from the Sun being received by and reflected from Earth's surface is related to the role of the atmosphere in maintaining conditions suitable for life.

Communicate Your Results

Display data from your initial observations to compare your findings with your classmates' findings. Explain your hypothesis, experiment results, and conclusions to the class.

Inquiry Extension

What could you add to this investigation to show how cloud cover changes the amount of radiation that will reach Earth's surfaces? Design a study that could test the effect of cloud cover on radiation passing through Earth's atmosphere. How could you include a way to show that clouds also reflect radiant energy from the Sun?

Lab Tips

☑ If possible, use leaves, straw, shaved ice, and other natural materials to test your hypothesis.

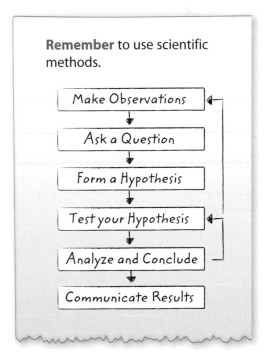

Remember to use scientific methods.

- Make Observations
- Ask a Question
- Form a Hypothesis
- Test your Hypothesis
- Analyze and Conclude
- Communicate Results

Chapter 12 Study Guide

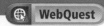

THE BIG IDEA The gases in Earth's atmosphere, some of which are needed by organisms to survive, affect Earth's temperature and the transfer of thermal energy to the atmosphere.

Key Concepts Summary

Vocabulary

Lesson 1: Describing Earth's Atmosphere

- Earth's **atmosphere** formed as Earth cooled and chemical and biological processes took place.
- Earth's atmosphere consists of nitrogen, oxygen, and a small amount of other gases, such as CO_2 and **water vapor**.
- The atmospheric layers are the **troposphere**, the **stratosphere**, the mesosphere, the thermosphere, and the exosphere.
- Air pressure decreases as altitude increases. Temperature either increases or decreases as altitude increases, depending on the layer of the atmosphere.

atmosphere p. 409
water vapor p. 410
troposphere p. 412
stratosphere p. 412
ozone layer p. 412
ionosphere p. 413

Lesson 2: Energy Transfer in the Atmosphere

- The Sun's energy is transferred to Earth's surface and the atmosphere through **radiation**, **conduction**, **convection**, and latent heat.
- Air circulation patterns are created by convection currents.

radiation p. 418
conduction p. 421
convection p. 421
stability p. 422
temperature inversion p. 423

Lesson 3: Air Currents

- Uneven heating of Earth's surface creates pressure differences. **Wind** is the movement of air from areas of high pressure to areas of low pressure.
- Air currents curve to the right or to the left due to the Coriolis effect.
- The main wind belts on Earth are the **trade winds**, the **westerlies**, and the **polar easterlies**.

wind p. 427
trade winds p. 429
westerlies p. 429
polar easterlies p. 429
jet stream p. 429
sea breeze p. 430
land breeze p. 430

Lesson 4: Air Quality

- Some human activities release pollution into the air.
- Air quality standards are monitored for the health of organisms and to determine if anti-pollution efforts are successful.

air pollution p. 434
acid precipitation p. 435
photochemical smog p. 435
particulate matter p. 436

Study Guide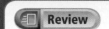

- Personal Tutor
- Vocabulary eGames
- Vocabulary eFlashcards

Chapter Project

Assemble your lesson Foldables® as shown to make a Chapter Project. Use the project to review what you have learned in this chapter.

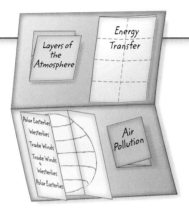

Use Vocabulary

1. Radio waves travel long distances by bouncing off electrically charged particles in the _____.

2. The Sun's thermal energy is transferred to Earth through space by _____.

3. Rising currents of warm air transfer energy from Earth to the atmosphere through _____.

4. A narrow band of winds located near the top of the troposphere is a(n) _____.

5. _____ are steady winds that flow from east to west between 30°N latitude and 30°S latitude.

6. In large urban areas, _____ forms when pollutants in the air interact with sunlight.

7. A mixture of dust, acids, and other chemicals that can be hazardous to human health is called _____.

Link Vocabulary and Key Concepts

Concepts in Motion — Interactive Concept Map

Copy this concept map, and then use vocabulary terms from the previous page to complete the concept map.

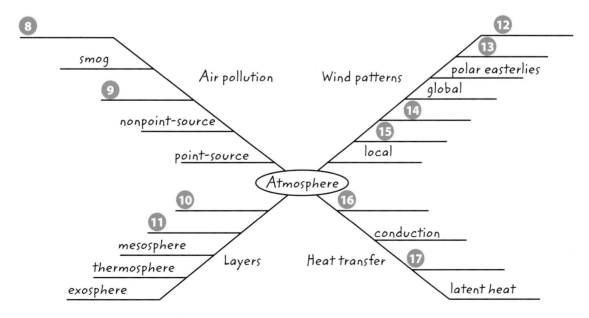

Chapter 12 Review

Understand Key Concepts

1. Air pressure is greatest
   A. at a mountain base.
   B. on a mountain top.
   C. in the stratosphere.
   D. in the ionosphere.

2. In which layer of the atmosphere is the ozone layer found?
   A. troposphere
   B. stratosphere
   C. mesosphere
   D. thermosphere

Use the image below to answer question 3.

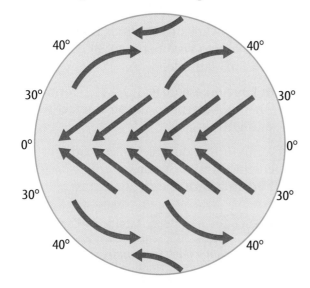

3. This diagram represents the atmosphere's
   A. air masses.
   B. global wind belts.
   C. inversions.
   D. particulate motion.

4. The Sun's energy
   A. is completely absorbed by the atmosphere.
   B. is completely reflected by the atmosphere.
   C. is in the form of latent heat.
   D. is transferred to the atmosphere after warming Earth.

5. Which type of energy is emitted from Earth to the atmosphere?
   A. ultraviolet radiation
   B. visible radiation
   C. infrared radiation
   D. aurora borealis

6. Which is a narrow band of high winds located near the top of the troposphere?
   A. polar easterly
   B. a jet stream
   C. a sea breeze
   D. a trade wind

7. Which helps protect people, animals, plants, and buildings from the harmful effects of air pollution?
   A. primary pollutants
   B. secondary pollutants
   C. ozone layer
   D. air quality standards

Use the photo below to answer question 8.

8. This photo shows a potential source of
   A. ultraviolet radiation.
   B. indoor air pollution.
   C. radon.
   D. smog.

Chapter Review

Assessment
Online Test Practice

Critical Thinking

9 **Predict** how atmospheric carbon dioxide levels might change if more trees were planted on Earth. Explain your prediction.

10 **Compare** visible and infrared radiation.

11 **Assess** whether your home is heated by conduction or convection.

12 **Sequence** how the unequal heating of Earth's surface leads to the formation of wind.

13 **Evaluate** whether a sea breeze could occur at night.

14 **Interpret Graphics** What are the top three sources of particulate matter in the atmosphere? What could you do to reduce particulate matter from any of the sources shown here?

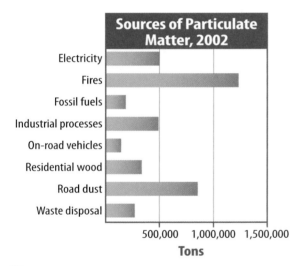

15 **Diagram** how acid precipitation forms. Include possible sources of sulfur dioxide and nitrogen oxide and organisms that can be affected by acid precipitation.

Writing in Science

16 **Write** a paragraph explaining whether you think it would be possible to permanently pollute the atmosphere with particulate matter.

REVIEW THE BIG IDEA

17 Review the title of each lesson in the chapter. List all of the characteristics and components of the troposphere and the stratosphere that affect life on Earth. Describe how life is impacted by each one.

18 Discuss how energy is transferred from the Sun throughout Earth's atmosphere.

Math Skills

Review Math Practice

Use Graphs

19 What was the percent change in energy use between 1996 and 1999?

20 What happened to energy use between 1999 and 2000?

21 What was the total percentage change between vehicle miles traveled and air pollution from 1970 to 2000?

Standardized Test Practice

Record your answers on the answer sheet provided by your teacher or on a sheet of paper.

Multiple Choice

1. What causes the phenomenon known as a mountain wave?
   - A radiation imbalance
   - B rising and sinking air
   - C temperature inversion
   - D the greenhouse effect

Use the diagram below to answer question 2.

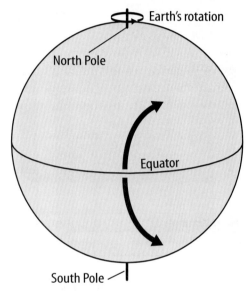

2. What phenomenon does the diagram above illustrate?
   - A radiation balance
   - B temperature inversion
   - C the Coriolis effect
   - D the greenhouse effect

3. Which do scientists call greenhouse gases?
   - A carbon dioxide, hydrogen, nitrogen
   - B carbon dioxide, methane gas, water vapor
   - C carbon monoxide, oxygen, argon
   - D carbon monoxide, ozone, radon

4. In which direction does moving air appear to turn in the northern hemisphere?
   - A down
   - B up
   - C right
   - D left

Use the diagram below to answer question 5.

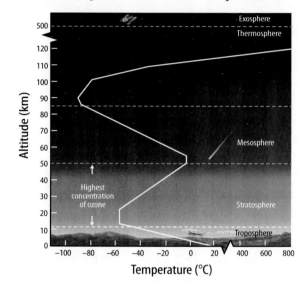

5. Which layer of the atmosphere has the widest range of temperatures?
   - A mesosphere
   - B stratosphere
   - C thermosphere
   - D troposphere

6. Which was the main component of Earth's original atmosphere?
   - A carbon dioxide
   - B nitrogen
   - C oxygen
   - D water vapor

446 • Chapter 12 Standardized Test Practice

Standardized Test Practice

7 Which is the primary cause of the global wind patterns on Earth?
  A ice cap melting
  B uneven heating
  C weather changing
  D waves breaking

Use the diagram below to answer question 8.

Energy Transfer Methods

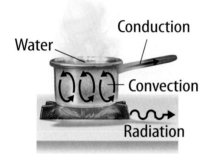

8 In the diagram above, which transfers thermal energy in the same way the Sun's energy is transferred to Earth?
  A the boiling water
  B the burner flame
  C the hot handle
  D the rising steam

9 Which substance in the air of U.S. cities has decreased least since the Clean Air Act began?
  A carbon monoxide
  B ground-level ozone
  C particulate matter
  D sulfur dioxide

Constructed Response

Use the table below to answer questions 10 and 11.

Layer	Significant Fact

10 In the table above, list in order the layers of Earth's atmosphere from lowest to highest. Provide one significant fact about each layer.

11 Explain how the first four atmospheric layers are important to life on Earth.

Use the table below to answer question 12.

Heat Transfer	Explanation
Conduction	
Convection	
Latent heat	
Radiation	

12 Complete the table to explain how heat energy transfers from the Sun to Earth and its atmosphere.

13 What are temperature inversions? How do they form? What is the relationship between temperature inversions and air pollution?

NEED EXTRA HELP?													
If You Missed Question...	1	2	3	4	5	6	7	8	9	10	11	12	13
Go to Lesson...	2	3	2	3	1	1	3	2	4	1	1	2	2,4

Chapter 13

Weather

THE BIG IDEA How do scientists describe and predict weather?

Inquiry Is this a record snowfall?

Buffalo, New York, is famous for its snowstorms, averaging 3 m of snow each year. Other areas of the world might only get a few centimeters of snow a year. In some parts of the world, it never snows.

- Why do some areas get less snow than others?
- How do scientists describe and predict weather?

Get Ready to Read

What do you think?

Before you read, decide if you agree or disagree with each of these statements. As you read this chapter, see if you change your mind about any of the statements.

1. Weather is the long-term average of atmospheric patterns of an area.
2. All clouds are at the same altitude within the atmosphere.
3. Precipitation often occurs at the boundaries of large air masses.
4. There are no safety precautions for severe weather, such as tornadoes and hurricanes.
5. Weather variables are measured every day at locations around the world.
6. Modern weather forecasts are done using computers.

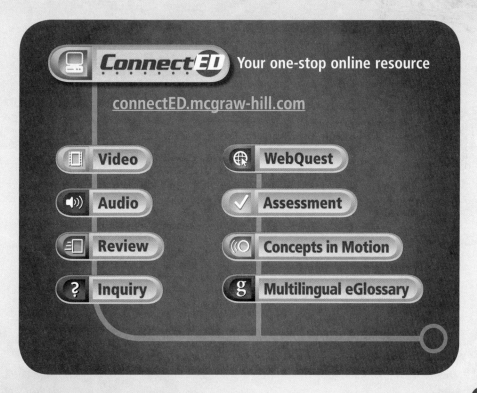

ConnectED — Your one-stop online resource

connectED.mcgraw-hill.com

- Video
- WebQuest
- Audio
- Assessment
- Review
- Concepts in Motion
- Inquiry
- Multilingual eGlossary

Lesson 1

Reading Guide

Key Concepts
ESSENTIAL QUESTIONS

- What is weather?
- What variables are used to describe weather?
- How is weather related to the water cycle?

Vocabulary

weather p. 451
air pressure p. 452
humidity p. 452
relative humidity p. 453
dew point p. 453
precipitation p. 455
water cycle p. 455

 Multilingual eGlossary

 Video

- BrainPOP®
- Science Video

Describing Weather

Inquiry Why are clouds different?

If you look closely at the photo, you'll see that there are different types of clouds in the sky. How do clouds form? If all clouds consist of water droplets and ice crystals, why do they look different? Are clouds weather?

Inquiry Launch Lab

15 minutes

Can you make clouds in a bag?

When water vapor in the atmosphere cools, it condenses. The resulting water droplets make up clouds.

1. Read and complete a lab safety form.
2. Half-fill a **500-mL beaker** with **ice** and **cold water.**
3. Pour 125 mL of **warm water** into a **resealable plastic bag** and seal the bag.
4. Carefully lower the bag into the ice water. Record your observations in your Science Journal.

Think About This

1. What did you observe when the warm water in the bag was put into the beaker?
2. What explanation can you give for what happened?
3. **Key Concept** What could you see in the natural world that results from the same process?

What is weather?

Everybody talks about the weather. "Nice day, isn't it?" "How was the weather during your vacation?" Talking about weather is so common that we even use weather terms to describe unrelated topics. "That homework assignment was a breeze." Or "I'll take a rain check."

Weather is *the atmospheric conditions, along with short-term changes, of a certain place at a certain time.* If you have ever been caught in a rainstorm on what began as a sunny day, you know the weather can change quickly. Sometimes it changes in just a few hours. But other times your area might have the same sunny weather for several days in a row.

Weather Variables

Perhaps some of the first things that come to mind when you think about weather are temperature and rainfall. As you dress in the morning, you need to know what the temperature will be throughout the day to help you decide what to wear. If it is raining, you might cancel your picnic.

Temperature and rainfall are just two of the variables used to describe weather. Meteorologists, scientists who study and predict weather, use several specific variables that describe a variety of atmospheric conditions. These variables include air temperature, air pressure, wind speed and direction, humidity, cloud coverage, and precipitation.

Key Concept Check What is weather?

REVIEW VOCABULARY
variable
a quantity that can change

Lesson 1
EXPLORE
451

REVIEW VOCABULARY

kinetic energy
energy an object has due to its motion

Air Temperature

The measure of the average kinetic energy of molecules in the air is air temperature. When the temperature is high, molecules have a high kinetic energy. Therefore, molecules in warm air move faster than molecules in cold air. Air temperatures vary with time of day, season, location, and altitude.

Air Pressure

The force that a column of air applies on the air or a surface below it is called **air pressure.** Study **Figure 1.** Is air pressure at Earth's surface more or less than air pressure at the top of the atmosphere? Air pressure decreases as altitude increases. Therefore, air pressure is greater at low altitudes than at high altitudes.

You might have heard the term *barometric pressure* during a weather forecast. Barometric pressure refers to air pressure. Air pressure is measured with an instrument called a barometer, shown in **Figure 2.** Air pressure is typically measured in millibars (mb). Knowing the barometric pressure of different areas helps meteorologists predict the weather.

 Reading Check What instrument measures air pressure?

Wind

As air moves from areas of high pressure to areas of low pressure, it creates wind. Wind direction is the direction from which the wind is blowing. For example, winds that blow from west to east are called westerlies. Meteorologists measure wind speed using an instrument called an anemometer (a nuh MAH muh tur). An anemometer is also shown in **Figure 2.**

Humidity

The amount of water vapor in the air is called **humidity** (hyew MIH duh tee). Humidity can be measured in grams of water per cubic meter of air (g/m^3). When the humidity is high, there is more water vapor in the air. On a day with high humidity, your skin might feel sticky, and sweat might not evaporate from your skin as quickly.

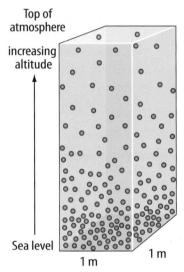

Figure 1 Increasing air pressure comes from having more molecules overhead.

Visual Check What happens to air pressure as altitude decreases?

Figure 2 Barometers, left, and anemometers, right, are used to measure weather variables.

Relative Humidity

Think about how a sponge can absorb water. At some point, it becomes full and cannot absorb any more water. In the same way, air can only contain a certain amount of water vapor. When air is saturated, it contains as much water vapor as possible. Temperature determines the maximum amount of water vapor air can contain. Warm air can contain more water vapor than cold air. *The amount of water vapor present in the air compared to the maximum amount of water vapor the air could contain at that temperature is called* **relative humidity.**

Relative humidity is measured using an instrument called a psychrometer and is given as a percent. For example, air with a relative humidity of 100 percent cannot contain any more moisture and dew or rain will form. Air that contains only half the water vapor it could hold has a relative humidity of 50 percent.

 Reading Check Compare and contrast humidity and relative humidity.

Dew Point

When a sponge becomes saturated with water, the water starts to drip from the sponge. Similarly, when air becomes saturated with water vapor, the water vapor will condense and form water droplets. When air near the ground becomes saturated, the water vapor in air will condense to a liquid. If the temperature is above 0°C, dew forms. If the temperature is below 0°C, ice crystals, or frost, form. Higher in the atmosphere clouds form. The graph in **Figure 3** shows the total amount of water vapor that air can contain at different temperatures.

When the temperature decreases, the air can hold less moisture. As you just read, the air becomes saturated, condensation occurs, and dew forms. *The temperature at which air is saturated and condensation can occur is called the* **dew point.**

Inquiry MiniLab   20 minutes

When will dew form?
The relative humidity on a summer day is 80 percent. The temperature is 35°C. Will the dew point be reached if the temperature drops to 25°C later in the evening? Use **Figure 3** below to find the amount of water vapor needed for saturation at each temperature.

1. Calculate the amount of water vapor in air that is 35°C and has 80 percent relative humidity. (Hint: multiply the amount of water vapor air can contain at 35°C by the percent of relative humidity.)

2. At 25°C, air can hold 2.2 g/cm^3 of water vapor. If your answer from step 1 is less than 2.2 g/cm^3, the dew point is not reached and dew will not form. If the number is greater, dew will form.

Analyze and Conclude
Key Concept After the Sun rises in the morning the air's temperature increases. How does the relative humidity change after sunrise? What does the line represent?

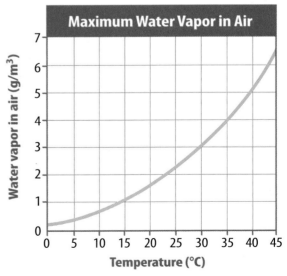

Figure 3 As air temperature increases, the air can contain more water vapor.

Figure 4 Clouds have different shapes and can be found at different altitudes.

Stratus clouds
- flat, white, and layered
- altitude up to 2,000 m

Cumulus clouds
- fluffy, heaped, or piled up
- 2,000 to 6,000 m altitude

Cirrus clouds
- wispy
- above 6,000 m

WORD ORIGIN

precipitation
from Latin *praecipitationem*, means "act or fact of falling headlong"

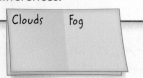

Make a horizontal two-tab book and label the tabs as illustrated. Use it to collect information on clouds and fog. Find similarities and differences.

Clouds and Fog

When you exhale outside on a cold winter day, you can see the water vapor in your breath condense into a foggy cloud in front of your face. This also happens when warm air containing water vapor cools as it rises in the atmosphere. When the cooling air reaches its dew point, water vapor condenses on small particles in the air and forms droplets. Surrounded by thousands of other droplets, these small droplets block and reflect light. This makes them visible as clouds.

Clouds are water droplets or ice crystals suspended in the atmosphere. Clouds can have different shapes and be present at different altitudes within the atmosphere. Different types of clouds are shown in **Figure 4.** Because we observe that clouds move, we recognize that water and thermal energy are transported from one location to another. Recall that clouds are also important in reflecting some of the Sun's incoming radiation.

A cloud that forms near Earth's surface is called fog. Fog is a suspension of water droplets or ice crystals close to or at Earth's surface. Fog reduces visibility, the distance a person can see into the atmosphere.

Reading Check What is fog?

Precipitation

Recall that droplets in clouds form around small solid particles in the atmosphere. These particles might be dust, salt, or smoke. Precipitation occurs when cloud droplets combine and become large enough to fall back to Earth's surface. **Precipitation** is *water, in liquid or solid form, that falls from the atmosphere.* Examples of precipitation—rain, snow, sleet, and hail—are shown in **Figure 5.**

Rain is precipitation that reaches Earth's surface as droplets of water. Snow is precipitation that reaches Earth's surface as solid, frozen crystals of water. Sleet may originate as snow. The snow melts as it falls through a layer of warm air and refreezes when it passes through a layer of below-freezing air. Other times it is just freezing rain. Hail reaches Earth's surface as large pellets of ice. Hail starts as a small piece of ice that is repeatedly lifted and dropped by an updraft within a cloud. A layer of ice is added with each lifting. When it finally becomes too heavy for the updraft to lift, it falls to Earth.

 Key Concept Check What variables are used to describe weather?

The Water Cycle

Precipitation is an important process in the water cycle. Evaporation and condensation are phase changes that are also important to the water cycle. *The* **water cycle** *is the series of natural processes by which water continually moves among oceans, land, and the atmosphere.* As illustrated in **Figure 6,** most water vapor enters the atmosphere when water at the ocean's surface is heated and evaporates. Water vapor cools as it rises in the atmosphere and condenses back into a liquid. Eventually, droplets of liquid and solid water form clouds. Clouds produce precipitation, which falls to Earth's surface and later evaporates, continuing the cycle.

 Key Concept Check How is weather related to the water cycle?

Types of Precipitation

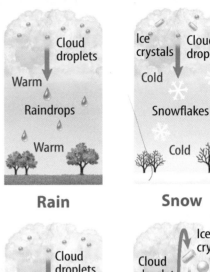
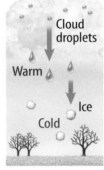
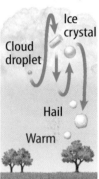

▲ **Figure 5** Rain, snow, sleet, and hail are forms of precipitation.

 Visual Check What is the difference between snow and sleet?

The Water Cycle

Figure 6 The Sun's energy powers the water cycle, which is the continual movement of water between the ocean, the land, and the atmosphere.

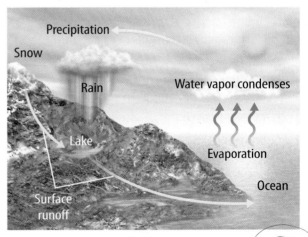

Lesson 1 Review

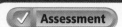

Visual Summary

Weather is the atmospheric conditions, along with short-term changes, of a certain place at a certain time.

Meteorologists use weather variables to describe atmospheric conditions.

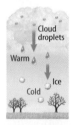

Forms of precipitation include rain, sleet, snow, and hail.

FOLDABLES

Use your lesson Foldable to review the lesson. Save your Foldable for the project at the end of the chapter.

What do you think NOW?

You first read the statements below at the beginning of the chapter.

1. Weather is the long-term average of atmospheric patterns of an area.

2. All clouds are at the same altitude within the atmosphere.

Did you change your mind about whether you agree or disagree with the statements? Rewrite any false statements to make them true.

Use Vocabulary

1 Define *humidity* in your own words.

2 Use the term *precipitation* in a sentence.

3 _____ is the pressure that a column of air exerts on the surface below it.

Understand Key Concepts

4 Which is NOT a standard weather variable?
   A. air pressure
   B. moon phase
   C. temperature
   D. wind speed

5 Identify and describe the different variables used to describe weather.

6 Relate humidity to cloud formation.

7 Describe how processes in the water cycle are related to weather.

Interpret Graphics

8 Identify Which type of precipitation is shown in the diagram below? How does this precipitation form?

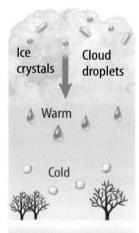

Critical Thinking

9 Analyze Why would your ears pop if you climbed a tall mountain?

10 Differentiate among cloud formation, fog formation, and dew point.

456 Chapter 13
EVALUATE

SCIENCE & SOCIETY

AMERICAN MUSEUM OF NATURAL HISTORY

Flooding caused widespread devastation in New Orleans, a city that lies below sea level. The storm surge broke through levees that had protected the city.

Is there a link between hurricanes and global warming?

Scientists worry that hurricanes might be getting bigger and happening more often.

On August 29, 2005, Hurricane Katrina roared through New Orleans, Louisiana. The storm destroyed homes and broke through levees, flooding most of the low-lying city. In the wake of the disaster, many wondered whether global warming was responsible. If warm oceans are the fuel for hurricanes, could rising temperatures cause stronger or more frequent hurricanes?

Climate scientists have several ways to investigate this question. They examine past hurricane activity, sea surface temperature, and other climate data. They compare these different types of data and look for patterns. Based on the laws of physics, they put climate and hurricane data into equations. A computer solves these equations and makes computer models. Scientists analyze the models to see whether there is a connection between hurricane activity and different climate variables.

What have scientists learned? So far they have not found a link between warming oceans and the frequency of hurricanes. However, they have found a connection between warming oceans and hurricane strength. Models suggest that rising ocean temperatures might create more destructive hurricanes with stronger winds and more rainfall.

The warm waters of the Gulf of Mexico fueled Hurricane Katrina as it spun toward Louisiana.

But global warming is not the only cause of warming oceans. As the ocean circulates, it goes through cycles of warming and cooling. Data show that the Atlantic Ocean has been in a warming phase for the past few decades.

Whether due to global warming or natural cycles, ocean temperatures are expected to rise even more in coming years. While rising ocean temperatures might not produce more hurricanes, climate research shows they could produce more powerful hurricanes. Perhaps the better question is not what caused Hurricane Katrina, but how we can prepare for equal-strength or more destructive hurricanes in the future.

It's Your Turn

DIAGRAM With a partner, create a storyboard with each frame showing one step in hurricane formation. Label your drawings. Share your storyboard with the class.

Lesson 1 EXTEND

Lesson 2

Weather Patterns

Reading Guide

Key Concepts
ESSENTIAL QUESTIONS

- What are two types of pressure systems?
- What drives weather patterns?
- Why is it useful to understand weather patterns?
- What are some examples of severe weather?

Vocabulary
high-pressure system p. 459
low-pressure system p. 459
air mass p. 460
front p. 462
tornado p. 465
hurricane p. 466
blizzard p. 467

 Multilingual eGlossary

 Video
What's Science Got to do With It?

 What caused this flooding?

Surging waves and rain from Hurricane Katrina caused flooding in New Orleans, Louisiana. Why are flooding and other types of severe weather dangerous? How does severe weather form?

458 • Chapter 13
ENGAGE

Launch Lab

10 minutes

How can temperature affect pressure?

Air molecules that have low energy can be packed closely together. As energy is added to the molecules they begin to move and bump into one another.

1. Read and complete a lab safety form.
2. Close a **resealable plastic bag** except for a small opening. Insert a **straw** through the opening and blow air into the bag until it is as firm as possible. Remove the straw and quickly seal the bag.
3. Submerge the bag in a **container** of **ice water** and hold it there for 2 minutes. Record your observations in your Science Journal.
4. Remove the bag from the ice water and submerge it in **warm water** for 2 minutes. Record your observations.

Think About This

1. What do the results tell you about the movement of air molecules in cold air and in warm air?
2. **Key Concept** What property of the air is demonstrated in this activity?

Pressure Systems

Weather is often associated with pressure systems. Recall that air pressure is the weight of the molecules in a large mass of air. When air molecules are cool, they are closer together than when they are warm. Cool air masses have high pressure, or more weight. Warm air masses have low pressure.

A **high-pressure system**, shown in **Figure 7**, is a large body of circulating air with high pressure at its center and lower pressure outside of the system. Because air moves from high pressure to low pressure, the air inside the system moves away from the center. Dense air sinks, bringing clear skies and fair weather.

A **low-pressure system**, also shown in **Figure 7**, is a large body of circulating air with low pressure at its center and higher pressure outside of the system. This causes air inside the low pressure system to rise. The rising air cools and the water vapor condenses, forming clouds and sometimes precipitation—rain or snow.

Key Concept Check Compare and contrast two types of pressure systems.

Figure 7 Air moving from areas of high pressure to areas of low pressure is called wind.

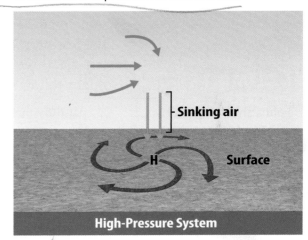

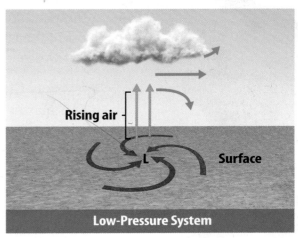

Lesson 2

459

EXPLORE

Air Masses

Figure 8 Five main air masses impact climate across North America.

Visual Check Where does continental polar air come from?

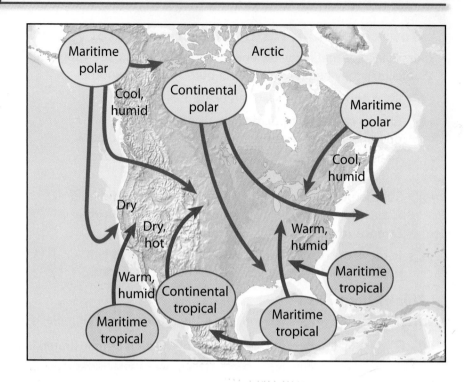

Air Masses

Have you ever noticed that the weather sometimes stays the same for several days in a row? For example, during winter in the northern United States, extremely cold temperatures often last for three or four days in a row. Afterward, several days might follow with warmer temperatures and snow showers.

Air masses are responsible for this pattern. **Air masses** *are large bodies of air that have uniform temperature, humidity, and pressure.* An air mass forms when a large high pressure system lingers over an area for several days. As a high pressure system comes in contact with Earth, the air in the system takes on the temperature and moisture characteristics of the surface below it.

Like high- and low-pressure systems, air masses can extend for a thousand kilometers or more. Sometimes one air mass covers most of the United States. Examples of the main air masses that affect weather in the United States are shown in **Figure 8**.

Air Mass Classification

Air masses are classified by their temperature and moisture characteristics. Air masses that form over land are referred to as continental air masses. Those that form over water are referred to as maritime masses. Warm air masses that form in the equatorial regions are called tropical. Those that form in cold regions are called polar. Air masses near the poles, over the coldest regions of the globe, are called arctic and antarctic air masses.

FOLDABLES

Fold a sheet of paper into thirds along the long axis. Label the outside *Air Masses*. Make another fold about 2 inches from the long edge of the paper to make a three-column chart. Label as shown.

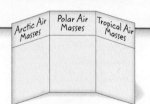

Arctic Air Masses Forming over Siberia and the Arctic are arctic air masses. They contain bitterly cold, dry air. During winter, an arctic air mass can bring temperatures down to −40°C.

Continental Polar Air Masses Because land cannot transfer as much moisture to the air as oceans can, air masses that form over land are drier than air masses that form over oceans. Continental polar air masses are fast-moving and bring cold temperatures in winter and cool weather in summer. Find the continental polar air masses over Canada in **Figure 8**.

Maritime Polar Air Masses Forming over the northern Atlantic and Pacific Oceans, maritime polar air masses are cold and humid. They often bring cloudy, rainy weather.

Continental Tropical Air Masses Because they form in the tropics over dry, desert land, continental tropical air masses are hot and dry. They bring clear skies and high temperatures. Continental tropical air masses usually form during the summer.

Maritime Tropical Air Masses As shown in **Figure 8**, maritime tropical air masses form over the western Atlantic Ocean, the Gulf of Mexico, and the eastern Pacific Ocean. These moist air masses bring hot, humid air to the southeastern United States during summer. In winter, they can bring heavy snowfall.

Air masses can change as they move over the land and ocean. Warm, moist air can move over land and become cool and dry. Cold, dry air can move over water and become moist and warm.

Key Concept Check What drives weather patterns?

Math Skills

Conversions

To convert Fahrenheit (°F) units to Celsius (°C) units, use this equation:

$$°C = \frac{(°F - 32)}{1.8}$$

Convert 76°F to °C

1. Always perform the operation in parentheses first.

   (76°F − 32 = **44°F**)

2. Divide the answer from Step 1 by 1.8.

   $$\frac{44°F}{1.8} = 24°C$$

To convert °C to °F, follow the same steps using the following equation:

$$°F = (°C \times 1.8) + 32$$

Practice

1. Convert 86°F to °C.
2. Convert 37°C to °F.

- Review
- Math Practice
- Personal Tutor

Inquiry MiniLab   20 minutes

How can you observe air pressure?

Although air seems very light, air molecules do exert pressure. You can observe air pressure in action in this activity.

1. Read and complete a lab safety form.
2. Tightly cap the empty **plastic bottle**.
3. Place the bottle in a **bucket of ice** for 10 minutes. Record your observations in your Science Journal.

Analyze and Conclude

1. **Interpret** how air pressure affected the bottle.
2. **Key Concept** Discuss how changing air pressure in Earth's atmosphere affects other things on Earth, such as weather.

Fronts

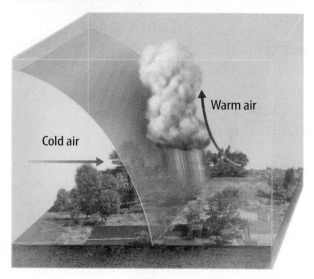 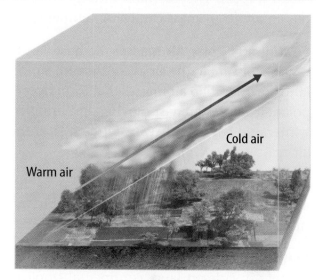

Cold   Warm

Figure 9 Certain types of fronts are associated with specific weather.

Visual Check Describe the difference between a cold front and a warm front.

SCIENCE USE V. COMMON USE

front

Science Use a boundary between two air masses

Common Use the foremost part or surface of something

Fronts

In 1918, Norwegian meteorologist Jacob Bjerknes (BYURK nehs) and his coworkers were busy developing a new method for forecasting the weather. Bjerknes noticed that specific types of weather occur at the boundaries between different air masses. Because he was trained in the army, Bjerknes used a military term to describe this boundary—front.

A military front is the boundary between opposing armies in a battle. *A weather* **front**, *however, is a boundary between two air masses.* Drastic weather changes often occur at fronts. As wind carries an air mass away from the area where it formed, the air mass will eventually collide with another air mass. Changes in temperature, humidity, cloud types, wind, and precipitation are common at fronts.

Cold Fronts

When a colder air mass moves toward a warmer air mass, a cold front forms, as shown in **Figure 9.** The cold air, which is denser than the warm air, pushes underneath the warm air mass. The warm air rises and cools. Water vapor in the air condenses and clouds form. Showers and thunderstorms often form along cold fronts. It is common for temperatures to decrease as much as 10°C when a cold front passes through. The wind becomes gusty and changes direction. In many cases, cold fronts give rise to severe storms.

Reading Check What types of weather are associated with cold fronts?

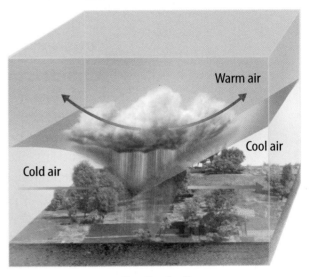

Stationary   Occluded

Warm Fronts

As shown in **Figure 9,** a warm front forms when less dense, warmer air moves toward colder, denser air. The warm air rises as it glides above the cold air mass. When water vapor in the warm air condenses, it creates a wide blanket of clouds. These clouds often bring steady rain or snow for several hours or even days. A warm front not only brings warmer temperatures, but it also causes the wind to shift directions.

Both a cold front and a warm front form at the edge of an approaching air mass. Because air masses are large, the movement of fronts is used to make weather forecasts. When a cold front passes through your area, temperatures will remain low for the next few days. When a warm front arrives, the weather will become warmer and more humid.

Stationary and Occluded Fronts

Sometimes an approaching front will stall for several days with warm air on one side of it and cold air on the other side. When the boundary between two air masses stalls, the front is called a stationary front. Study the stationary front shown in **Figure 9.** Cloudy skies and light rain are found along stationary fronts.

Cold fronts move faster than warm fronts. When a fast-moving cold front catches up with a slow-moving warm front, an occluded or blocked front forms. Occluded fronts, shown in **Figure 9,** usually bring precipitation.

Key Concept Check Why is it useful to understand weather patterns associated with fronts?

Severe Weather

Some weather events can cause major damage, injuries, and death. These events, such as thunderstorms, tornadoes, hurricanes, and blizzards, are called severe weather.

Thunderstorms

Also known as electrical storms because of their lightning, thunderstorms have warm temperatures, moisture, and rising air, which may be supplied by a low-pressure system. When these conditions occur, a cumulus cloud can grow into a 10-km-tall thundercloud, or cumulonimbus cloud, in as little as 30 minutes.

A typical thunderstorm has a three-stage life cycle, shown in **Figure 10**. The cumulus stage is dominated by cloud formation and updrafts. Updrafts are air currents moving vertically away from the ground. After the cumulus cloud has been created, downdrafts begin to appear. Downdrafts are air currents moving vertically toward the ground. In the mature stage, heavy winds, rain, and lightning dominate the area. Within 30 minutes of reaching the mature stage, the thunderstorm begins to fade, or dissipate. In the dissipation stage, updrafts stop, winds die down, lightning ceases, and precipitation weakens.

Strong updrafts and downdrafts within a thunderstorm cause millions of tiny ice crystals to rise and sink, crashing into each other. This creates positively and negatively charged particles in the cloud. The difference in the charges of particles between the cloud and the charges of particles on the ground eventually creates electricity. This is seen as a bolt of lightning. Lightning can move from cloud to cloud, cloud to ground, or ground to cloud. It can heat the nearby air to more than 27,000°C. Air molecules near the bolt rapidly expand and then contract, creating the sound identified as thunder.

> **ACADEMIC VOCABULARY**
> **dominate**
> *(verb)* to exert the guiding influence on

Figure 10 Thunderstorms have distinct stages characterized by the direction in which air is moving.

Thunderstorms

Cumulus Stage

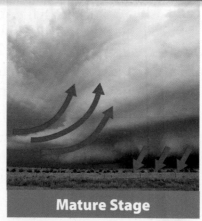
Mature Stage

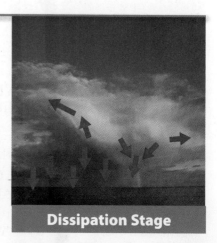
Dissipation Stage

Visual Check Describe what happens during each stage of a thunderstorm.

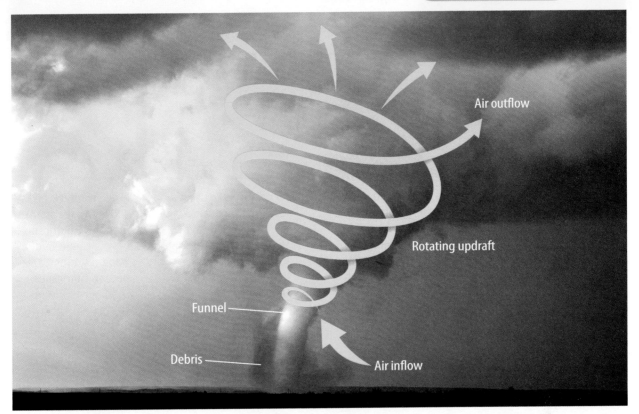

Figure 11 A funnel cloud forms when updrafts within a thunderstorm begin rotating.

Tornadoes

Perhaps you have seen photos of the damage from a tornado. *A tornado is a violent, whirling column of air in contact with the ground.* Most tornadoes have a diameter of several hundred meters. The largest tornadoes exceed 1,500 m in diameter. The intense, swirling winds within tornadoes can reach speeds of more than 400 km/h. These winds are strong enough to send cars, trees, and even entire houses flying through the air. Tornadoes usually last only a few minutes. More destructive tornadoes, however, can last for several hours.

Formation of Tornadoes When thunderstorm updrafts begin to rotate, as shown in **Figure 11**, tornadoes can form. Swirling winds spiral downward from the thunderstorm's base, creating a funnel cloud. When the funnel reaches the ground, it becomes a tornado. Although the swirling air is invisible, you can easily see the debris lifted by the tornado.

Reading Check How do tornadoes form?

Tornado Alley More tornadoes occur in the United States than anywhere else on Earth. The central United States, from Nebraska to Texas, experiences the most tornadoes. This area has been nicknamed Tornado Alley. In this area, cold air blowing southward from Canada frequently collides with warm, moist air moving northward from the Gulf of Mexico. These conditions are ideal for severe thunderstorms and tornadoes.

Classifying Tornadoes Dr. Ted Fujita developed a method for classifying tornadoes based on the damage they cause. On the modified Fujita intensity scale, F0 tornadoes cause light damage, breaking tree branches and damaging billboards. F1 though F4 tornadoes cause moderate to devastating damage, including tearing roofs from homes, derailing trains, and throwing vehicles in the air. F5 tornadoes cause incredible damage, such as demolishing concrete and steel buildings and pulling the bark from trees.

Hurricane Formation

Figure 12 Hurricanes consist of alternating bands of heavy precipitation and sinking air.

Hurricane Formation

1. As warm, moist air rises into the atmosphere, it cools, water vapor condenses, and clouds form. As more air rises, it creates an area of low pressure over the ocean.

2. As air continues to rise, a tropical depression forms. Tropical depressions bring thunderstorms with winds between 37–62 km/h.

3. Air continues to rise, rotating counterclockwise. The storm builds to a tropical storm with winds in excess of 63 km/h. It produces strong thunderstorms.

4. When winds exceed 119 km/h, the storm becomes a hurricane. Only one percent of tropical storms become hurricanes.

Inside a Hurricane

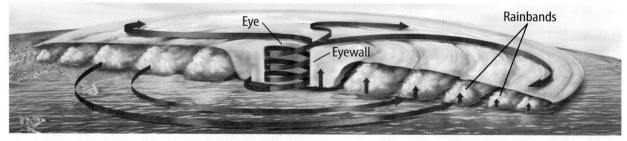

Visual Check How do hurricanes form?

WORD ORIGIN

hurricane
from Spanish *huracan*, means "tempest"

Hurricanes

An intense tropical storm with winds exceeding 119 km/h is a **hurricane.** Hurricanes are the most destructive storms on Earth. Like tornadoes, hurricanes have a circular shape with intense, swirling winds. However, hurricanes do not form over land. Hurricanes typically form in late summer over warm, tropical ocean water. **Figure 12** sequences the steps in hurricane formation. A typical hurricane is 480 km across, more than 150 thousand times larger than a tornado. At the center of a hurricane is the eye, an area of clear skies and light winds.

Damage from hurricanes occurs as a result of strong winds and flooding. While still out at sea, hurricanes create high waves that can flood coastal areas. As a hurricane crosses the coastline, or makes landfall, strong rains intensify and can flood and devastate entire areas. But once a hurricane moves over land or colder water, it loses its energy and dissipates.

In other parts of the world, these intense tropical storms have other names. In Asia, the same type of storm is called a typhoon. In Australia it is called a tropical cyclone.

Winter Storms

Not all severe weather occurs when temperatures are warm. Winter weather can also be severe. Snow and ice can make driving difficult and dangerous. When temperatures are close to freezing (0°C), rain can freeze when it hits the ground. Ice storms coat the ground, trees, and buildings with a layer of ice, as shown in **Figure 13**.

Figure 13 The weight of ice from freezing rain can cause trees, power lines, and other structures to break.

A **blizzard** *is a violent winter storm characterized by freezing temperatures, strong winds, and blowing snow.* During blizzards, blowing snow often reduces visibility to a few meters or even less. If you are outside during a blizzard, strong winds and very cold temperatures can rapidly cool exposed skin. Windchill, the combined cooling effect of cold temperature and wind on exposed skin, can cause frostbite and hypothermia (hi poh THER mee uh).

 Key Concept Check What are examples of severe weather?

Severe Weather Safety

To help keep people safe, the U.S. National Weather Service issues watches and warnings during severe weather events. A watch means that severe weather is possible. A warning means that severe weather is already occurring. Heeding severe weather watches and warnings is important and could save your life.

It is also important to know how to protect yourself during dangerous weather. During thunderstorms, you should stay inside if possible, and stay away from metal objects and electrical cords. If you are outside, stay away from water, high places and isolated trees. Dressing properly is important in all kinds of weather. When windchill temperatures are below −20°C you should dress in layers, keep your head and fingers covered, and limit your time outdoors.

Lesson 2 Review

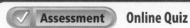

Visual Summary

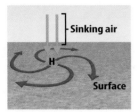

Low-pressure systems, high-pressure systems, and air masses all influence weather.

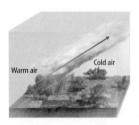

Weather often changes as a front passes through an area.

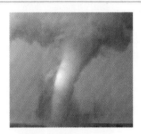

The National Weather Service issues warnings about severe weather such as thunderstorms, tornadoes, hurricanes, and blizzards.

FOLDABLES

Use your lesson Foldable to review the lesson. Save your Foldable for the project at the end of the chapter.

What do you think NOW?

You first read the statements below at the beginning of the chapter.

3. Precipitation often occurs at the boundaries of large air masses.

4. There are no safety precautions for severe weather, such as tornadoes and hurricanes.

Did you change your mind about whether you agree or disagree with the statements? Rewrite any false statements to make them true.

Use Vocabulary

1. **Distinguish** between an air mass and a front.

2. **Define** *low-pressure system* using your own words.

3. **Use the term** *high-pressure system* in a sentence.

Understand Key Concepts

4. Which air mass is humid and warm?
   A. continental polar
   B. continental tropical
   C. maritime polar
   D. maritime tropical

5. **Give an example** of cold-front weather.

6. **Compare and contrast** hurricanes and tornadoes.

7. **Explain** how thunderstorms form.

Interpret Graphics

8. **Compare and Contrast** Copy and fill in the graphic organizer below to compare and contrast high-pressure and low-pressure systems.

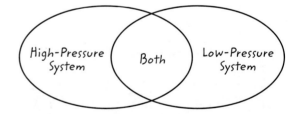

Critical Thinking

9. **Suggest** a reason that low-pressure systems are cloudy and rainy or snowy.

10. **Design** a pamphlet that contains tips on how to stay safe during different types of severe weather.

Math Skills

11. Convert 212°F to °C.

12. Convert 20°C to °F.

Inquiry Skill Practice — Recognize Cause and Effect — 30 minutes

Why does the weather change?

One day it is sunny, the next day it is pouring rain. If you look at only one location, the patterns that cause the weather to change are difficult to see. However, when you look on the large scale, the patterns become apparent.

Learn It

Recognizing cause and effect is an important part of science and conducting experiments. Scientists look for cause-and-effect relationships between variables. The maps below show the movement of fronts and pressure systems over a two-day period. What effect will these systems have on the weather as they move across the United States?

Try It

1. Examine the weather maps below. The thin black lines on each map represent areas where the barometric pressure is the same. The pressure is indicated by the number on the line. The center of a low- or high-pressure system is indicated by the word LOW or HIGH. Identify the location of low- and high- pressure systems on each map. Use the key below the maps to the identify the location of warm and cold fronts.

2. Find locations A, B, C, and where you live on the map. For each location, describe how the systems change positions over the two days.

3. What is the cause of and effect on precipitation and temperature at each location?

Apply It

4. The low-pressure system produced several tornadoes. Which location did they occur closest to? Explain.

5. The weather patterns generally move from west to east. Predict the weather on the third day for each location.

6. One day it is clear and sunny, but you notice that the pressure is less than it was the day before. What weather might be coming? Why?

7. **Key Concept** How does understanding weather patterns help make predicting the weather more accurate?

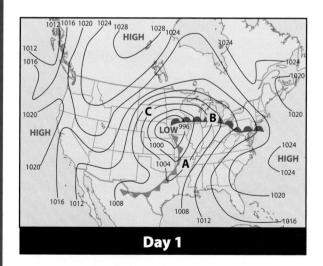
Day 1

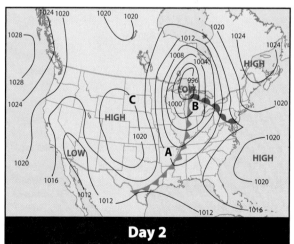
Day 2

Lesson 2 EXTEND — 469

Lesson 3

Weather Forecasts

Reading Guide

Key Concepts
ESSENTIAL QUESTIONS

- What instruments are used to measure weather variables?
- How are computer models used to predict the weather?

Vocabulary
surface report p. 471
upper-air report p. 471
Doppler radar p. 472
isobar p. 473
computer model p. 474

Multilingual eGlossary

Inquiry What's inside?

Information about weather variables is collected by the weather radar station shown here. Data, such as the amount of rain falling in a weather system, help meteorologists make accurate predictions about severe weather. What other instruments do meteorologists use to forecast weather? How do they collect and use data?

470 Chapter 13
ENGAGE

Launch Lab

10 minutes

Can you understand the weather report?

Weather reports use numbers and certain vocabulary terms to help you understand the weather conditions in a given area for a given time period. Listen to a weather report for your area. Can you record all the information reported?

1. In your Science Journal, make a list of data you would expect to hear in a weather report.
2. Listen carefully to a **recording of a weather report** and jot down numbers and measurements you hear next to those on your list.
3. Listen a second time and make adjustments to your original notes, such as adding more data, if necessary.
4. Listen a third time, then share the weather forecast as you heard it.

Think About This

1. What measurements were difficult for you to apply to understanding the weather report?
2. Why are so many different types of data needed to give a complete weather report?
3. List the instruments that might be used to collect each kind of data.
4. **Key Concept** Where do meteorologists obtain the data they use to make a weather forecast?

Measuring the Weather

Being a meteorologist is like being a doctor. Using specialized instruments and visual observations, the doctor first measures the condition of your body. The doctor later combines these measurements with his or her knowledge of medical science. The result is a forecast of your future health, such as, "You'll feel better in a few days if you rest and drink plenty of fluids."

Similarly, meteorologists, scientists who study weather, use specialized instruments to measure conditions in the atmosphere, as you read in Lesson 1. These instruments include thermometers to measure temperature, barometers to measure air pressure, psychrometers to measure relative humidity, and anemometers to measure wind speed.

Surface and Upper-Air Reports

A **surface report** *describes a set of weather measurements made on Earth's surface.* Weather variables are measured by a weather station—a collection of instruments that report temperature, air pressure, humidity, precipitation, and wind speed and direction. Cloud amounts and visibility are often measured by human observers.

An **upper-air report** *describes wind, temperature, and humidity conditions above Earth's surface.* These atmospheric conditions are measured by a radiosonde (RAY dee oh sahnd), a package of weather instruments carried many kilometers above the ground by a weather balloon. Radiosonde reports are made twice a day simultaneously at hundreds of locations around the world.

Satellite and Radar Images

Images taken from satellites orbiting about 35,000 km above Earth provide information about weather conditions on Earth. A visible light image, such as the one shown in **Figure 14,** shows white clouds over Earth. The infrared image, also shown in **Figure 14,** shows infrared energy in false color. The infrared energy comes from Earth and is stored in the atmosphere as thermal energy. Monitoring infrared energy provides information about cloud height and atmospheric temperature.

Figure 14 Meteorologists use visible light and infrared satellite images to identify fronts and air masses.

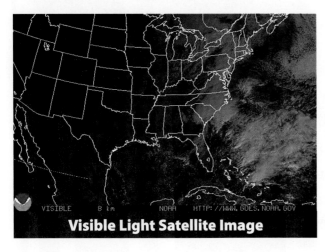
Visible Light Satellite Image

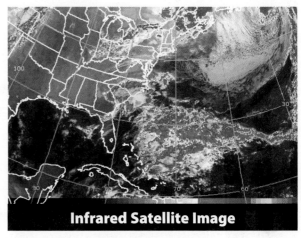
Infrared Satellite Image

Visual Check How is an infrared satellite image different from a visible light satellite image?

Radar measures precipitation when radio waves bounce off raindrops and snowflakes. **Doppler radar** *is a specialized type of radar that can detect precipitation as well as the movement of small particles, which can be used to approximate wind speed.* Because the movement of precipitation is caused by wind, Doppler radar can be used to estimate wind speed. This can be especially important during severe weather, such as tornadoes or thunderstorms.

Key Concept Check Identify the weather variables that radiosondes, infrared satellites, and Doppler radar measure.

Weather Maps

Every day, thousands of surface reports, upper-air reports, and satellite and radar observations are made around the world. Meteorologists have developed tools that help them simplify and understand this enormous amount of weather data.

FOLDABLES
Make a horizontal two-tab book and label the tabs as illustrated. Use it to collect information on satellite and radar images. Compare and contrast these information tools.

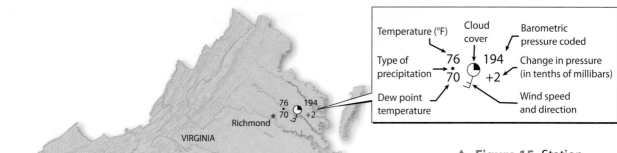

▲ Figure 15 Station models contain information about weather variables.

The Station Model

As shown in **Figure 15**, the station model diagram displays data from many different weather measurements for a particular location. It uses numbers and symbols to display data and observations from surface reports and upper-air reports.

Mapping Temperature and Pressure

In addition to station models, weather maps also have other symbols. For example, **isobars** *are lines that connect all places on a map where pressure has the same value.* Locate an isobar on the map in **Figure 16**. Isobars show the location of high- and low-pressure systems. Isobars also provide information about wind speed. Winds are strong when isobars are close together. Winds are weaker when isobars are farther apart.

In a similar way, isotherms (not shown) are lines that connect places with the same temperature. Isotherms show which areas are warm and which are cold. Fronts are represented as lines with symbols on them, as indicated in **Figure 16**.

Reading Check Compare isobars and isotherms.

WORD ORIGIN
isobar
from Greek *isos*, means "equal"; and *baros*, means "heavy"

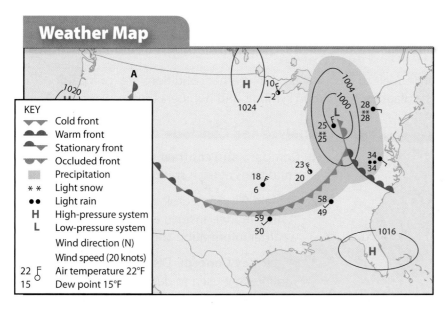

◀ Figure 16 Weather maps contain symbols that provide information about the weather.

Visual Check Which symbols represent high-pressure and low-pressure systems?

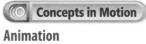
Animation

Lesson 3
EXPLAIN

Figure 17 Meteorologists analyze data from various sources—such as radar and computer models—in order to prepare weather forecasts.

Predicting the Weather

Modern weather forecasts are made with the help of computer models, such as the ones shown in **Figure 17**. **Computer models** *are detailed computer programs that solve a set of complex mathematical formulas.* The formulas predict what temperatures and winds might occur, when and where it will rain and snow, and what types of clouds will form.

Government meteorological offices also use computers and the Internet to exchange weather measurements continuously throughout the day. Weather maps are drawn and forecasts are made using computer models. Then, through television, radio, newspapers, and the Internet, the maps and forecasts are made available to the public.

 Key Concept Check How are computers used to predict the weather?

Inquiry MiniLab   20 minutes

How is weather represented on a map?

Meteorologists often use station models to record what the weather conditions are for a particular location. A station model is a diagram containing symbols and numbers that displays many different weather measurements.

Use the **station model legend** provided by your teacher to interpret the data in each station model shown here.

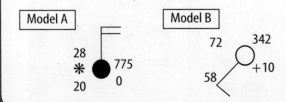

Analyze and Conclude

1. **Compare and contrast** the weather conditions at each station model.

2. **Explain** why meteorologists might use station models instead of reporting weather information another way.

3. **Key Concept** Discuss what variables are used to describe weather.

Lesson 3 Review

Assessment Online Quiz
Inquiry Virtual Lab

Visual Summary

Weather variables are measured by weather stations, radiosondes, satellites, and Doppler radar.

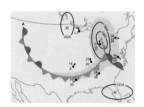

Weather maps contain information in the form of a station model, isobars and isotherms, and symbols for fronts and pressure systems.

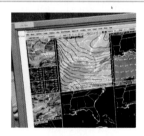

Meteorologists use computer models to help forecast the weather.

FOLDABLES

Use your lesson Foldable to review the lesson. Save your Foldable for the project at the end of the chapter.

What do you think NOW?

You first read the statements below at the beginning of the chapter.

5. Weather variables are measured every day at locations around the world.

6. Modern weather forecasts are done using computers.

Did you change your mind about whether you agree or disagree with the statements? Rewrite any false statements to make them true.

Use Vocabulary

1 **Define** *computer model* in your own words.

2 A line connecting places with the same pressure is called a(n) _____.

3 **Use the term** *surface report* in a sentence.

Understand Key Concepts

4 Which diagram shows surface weather measurements?
  A. an infrared satellite image
  B. an upper air chart
  C. a station model
  D. a visible light satellite image

5 **List** two ways that upper-air weather conditions are measured.

6 **Describe** how computers are used in weather forecasting.

7 **Distinguish** between isobars and isotherms.

Interpret Graphics

8 **Identify** Copy and fill in the graphic organizer below to identify the components of a surface map.

Symbol	Meaning
◠▼◠	
H	

Critical Thinking

9 **Suggest** ways to forecast the weather without using computers.

10 **Explain** why isobars and isotherms make it easier to understand a weather map.

Lesson 3
EVALUATE
475

Inquiry Lab

40 minutes

Can you predict the weather?

Materials

graph paper

local weather maps

outdoor thermometer

barometer

Weather forecasts are important—not just so you are dressed right when you leave the house, but also to help farmers know when to plant and harvest, to help cities know when to call in the snow plows, and to help officials know when and where to evacuate in advance of severe weather.

Ask a Question

Can you predict the weather?

Make Observations

1. Read and complete a lab safety form.
2. Collect weather data daily for a period of one week. Temperature and pressure should be recorded as a number, but precipitation, wind conditions, and cloud cover can be described in words. Make your observations at the same time each day.

3. Graph temperature in degrees and air pressure in millibars on the same sheet of paper, placing the graphs side by side, as shown on the next page. Beneath the graphs, for each day, add notes that describe precipitation, wind conditions, and cloud cover.

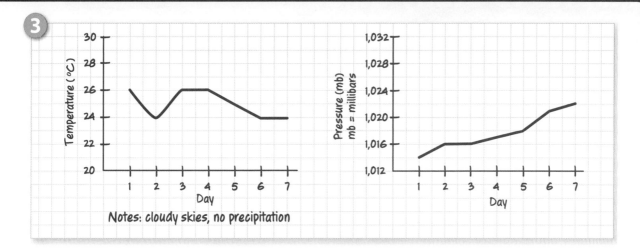

Form a Hypothesis

4. Examine your data and the weather maps. Look for factors that appear to be related. For example, your data might suggest that when the pressure decreases, clouds follow.

5. Find three sets of data pairs that seem to be related. Form three hypotheses, one for each set of data pairs.

Test Your Hypothesis

6. Look at your last day of data. Using your hypotheses, predict the weather for the next day.

7. Collect weather data the next day and evaluate your predictions.

8. Repeat steps 6 and 7 for at least two more days.

Analyze and Conclude

9. **Analyze** Compare your hypotheses with the results of your predictions. How successful were you? What additional information might have improved your predictions?

10. **The Big Idea** Scientists have more complex and sophisticated tools to help them predict their weather, but with fairly simple tools, you can make an educated guess. Write a one-paragraph summary of the data you collected and how you interpreted it to predict the weather.

Communicate Your Results

For each hypothesis you generated, make a small poster that states the hypothesis, shows a graph that supports it, and shows the results of your predictions. Write a concluding statement about the reliability of your hypothesis. Share your results with the class.

Investigate other forms of data you might collect and find out how they would help you to make a forecast. Try them out for a week and see if your ability to make predictions improves.

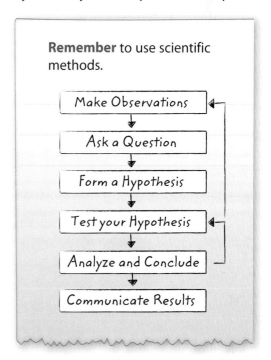

Chapter 13 Study Guide

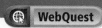

THE BIG IDEA Scientists use weather variables to describe weather and study weather systems. Scientists use computers to predict the weather.

Key Concepts Summary

Lesson 1: Describing Weather

- **Weather** is the atmospheric conditions, along with short-term changes, of a certain place at a certain time.
- Variables used to describe weather are air temperature, **air pressure**, wind, **humidity**, and **relative humidity**.
- The processes in the water cycle—evaporation, condensation, and **precipitation**—are all involved in the formation of different types of weather.

Vocabulary

weather p. 451
air pressure p. 452
humidity p. 452
relative humidity p. 453
dew point p. 453
precipitation p. 455
water cycle p. 455

Lesson 2: Weather Patterns

- **Low-pressure systems** and **high-pressure systems** are two systems that influence weather.
- Weather patterns are driven by the movement of **air masses.**
- Understanding weather patterns helps make weather forecasts more accurate.
- Severe weather includes thunderstorms, **tornadoes**, **hurricanes**, and **blizzards**.

high-pressure system p. 459
low-pressure system p. 459
air mass p. 460
front p. 462
tornado p. 465
hurricane p. 466
blizzard p. 467

Lesson 3: Weather Forecasts

- Thermometers, barometers, anemometers, radiosondes, satellites, and **Doppler radar** are used to measure weather variables.
- **Computer models** use complex mathematical formulas to predict temperature, wind, cloud formation, and precipitation.

surface report p. 471
upper-air report p. 471
Doppler radar p. 472
isobar p. 473
computer model p. 474

478 • Chapter 13 Study Guide

Study Guide

Review
- Personal Tutor
- Vocabulary eGames
- Vocabulary eFlashcards

FOLDABLES Chapter Project

Assemble your lesson Foldables as shown to make a Chapter Project. Use the project to review what you have learned in this chapter.

Use Vocabulary

1. The pressure that a column of air exerts on the area below it is called _____.

2. The amount of water vapor in the air is called _____.

3. The natural process in which water constantly moves among oceans, land, and the atmosphere is called the _____.

4. A(n) _____ is a boundary between two air masses.

5. At the center of a(n) _____, air rises and forms clouds and precipitation.

6. A continental polar _____ brings cold temperatures during winter.

7. When the same _____ passes through two locations on a weather map, both locations have the same pressure.

8. The humidity in the air compared to the amount air can hold is the _____.

Link Vocabulary and Key Concepts

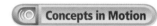 Interactive Concept Map

Copy this concept map, and then use vocabulary terms from the previous page to complete the concept map.

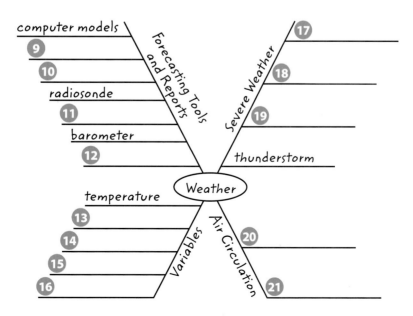

Chapter 13 Study Guide • **479**

Chapter 13 Review

Understand Key Concepts

1. Clouds form when water changes from
   A. gas to liquid.
   B. liquid to gas.
   C. solid to gas.
   D. solid to liquid.

2. Which type of precipitation reaches Earth's surface as large pellets of ice?
   A. hail
   B. rain
   C. sleet
   D. snow

3. Which of these sinking-air situations usually brings fair weather?
   A. air mass
   B. cold front
   C. high-pressure system
   D. low-pressure system

4. Which air mass contains cold, dry air?
   A. continental polar
   B. continental tropical
   C. maritime tropical
   D. maritime polar

5. Study the front below.

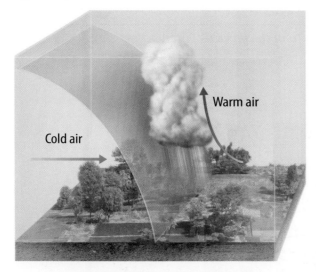

   How does this type of front form?
   A. A cold front overtakes a warm front.
   B. Cold air moves toward warmer air.
   C. The boundary between two fronts stalls.
   D. Warm air moves toward colder air.

6. Which is an intense tropical storm with winds exceeding 119 km/h?
   A. blizzard
   B. hurricane
   C. thunderstorm
   D. tornado

7. Which contains measurements of temperature, air pressure, humidity, precipitation, and wind speed and direction?
   A. a radar image
   B. a satellite image
   C. a surface report
   D. a weather station

8. What does Doppler radar measure?
   A. air pressure
   B. air temperature
   C. the rate at which air pressure changes
   D. the speed at which precipitation travels

9. Study the station model below.

   What is the temperature according to the station model?
   A. 3°F
   B. 55°F
   C. 81°F
   D. 138°F

10. Which describes cirrus clouds?
    A. flat, white, and layered
    B. fluffy, at middle altitudes
    C. heaped or piled up
    D. wispy, at high altitudes

11. Which instrument measures wind speed?
    A. anemometer
    B. barometer
    C. psychrometer
    D. thermometer

480 • Chapter 13 Review

Chapter Review

Assessment — Online Test Practice

Critical Thinking

12 Predict Suppose you are on a ship near the equator in the Atlantic Ocean. You notice that the barometric pressure is dropping. Predict what type of weather you might experience.

13 Compare a continental polar air mass with a maritime tropical air mass.

14 Assess why clouds usually form in the center of a low-pressure system.

15 Predict how maritime air masses would change if the oceans froze.

16 Compare two types of severe weather.

17 Interpret Graphics Identify the front on the weather map below. Predict the weather for areas along the front.

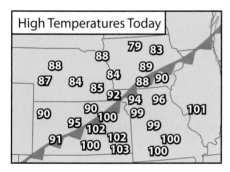

18 Assess the validity of the weather forecast: "Tomorrow's weather will be similar to today's weather."

19 Compare and contrast surface weather reports and upper-air reports. Why is it important for meterologists to monitor weather variables high above Earth's surface?

Writing in Science

20 Write a paragraph about the ways computers have improved weather forecasts. Be sure to include a topic sentence and a concluding sentence.

REVIEW THE BIG IDEA

21 Identify the instruments used to measure weather variables.

22 How do scientists use weather variables to describe and predict weather?

23 Describe the factors that influence weather.

24 Use the factors listed in question 23 to describe how a continental polar air mass can change to a maritime polar air mass.

Math Skills

Review — Math Practice

Use Conversions

25 Convert from Fahrenheit to Celsius.
  a. Convert 0°F to °C.
  b. Convert 104°F to °C.

26 Convert from Celsius to Fahrenheit.
  a. Convert 0°C to °F.
  b. Convert −40°C to °F.

27 The Kelvin scale of temperature measurement starts at zero and has the same unit size as Celsius degrees. Zero degrees Celsius is equal to 273 kelvin.

Convert 295 K to Fahrenheit.

Standardized Test Practice

Record your answers on the answer sheet provided by your teacher or on a sheet of paper.

Multiple Choice

1. Which measures the average kinetic energy of air molecules?
   A humidity
   B pressure
   C speed
   D temperature

Use the diagram below to answer question 2.

2. Which weather system does the above diagram illustrate?
   A high pressure
   B hurricane
   C low pressure
   D tornado

3. What causes weather to remain the same several days in a row?
   A air front
   B air mass
   C air pollution
   D air resistance

4. Which lists the stages of a thunderstorm in order?
   A cumulus, dissipation, mature
   B cumulus, mature, dissipation
   C dissipation, cumulus, mature
   D dissipation, mature, cumulus

5. What causes air to reach its dew point?
   A decreasing air currents
   B decreasing humidity
   C dropping air pressure
   D dropping temperatures

6. Which measures air pressure?
   A anemometer
   B barometer
   C psychrometer
   D thermometer

Use the diagram below to answer question 7.

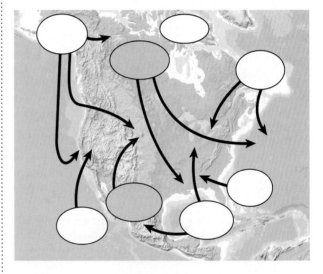

7. Which type of air masses do the shaded ovals in the diagram depict?
   A antarctic
   B arctic
   C continental
   D maritime

8. Which BEST expresses moisture saturation?
   A barometric pressure
   B relative humidity
   C weather front
   D wind direction

482 • Chapter 13 Standardized Test Practice

Standardized Test Practice

Use the diagram below to answer question 9.

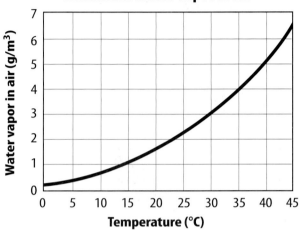

Maximum Water Vapor in Air

9 What happens to maximum moisture content when air temperatures increase from 15°C to 30°C?

   A increases from 1 to 2 g/m^3

   B increases from 1 to 3 g/m^3

   C increases from 2 to 3 g/m^3

   D increases from 2 to 4 g/m^3

10 When isobars are close together on a weather map,

   A cloud cover is extensive.

   B temperatures are high.

   C warm fronts prevail.

   D winds are strong.

11 Which provides energy for the water cycle?

   A air currents

   B Earth's core

   C ocean currents

   D the Sun

Constructed Response

Use the table below to answer question 12.

Weather Variable	Measurement

12 In the table above, list the variables weather scientists use to describe weather. Then describe the unit of measurement for each variable.

Use the diagram below to answer questions 13 and 14.

13 What does the diagram above depict?

14 Describe the weather conditions associated with the diagram.

15 How do weather fronts form?

NEED EXTRA HELP?															
If You Missed Question...	1	2	3	4	5	6	7	8	9	10	11	12	13	14	15
Go to Lesson...	1	2	2	2	1	1,3	2	1	1	3	1	1	2	2	2

Chapter 13 Standardized Test Practice • **483**

Chapter 14

Climate

THE BIG IDEA What is climate and how does it impact life on Earth?

Inquiry **What happened to this tree?**

Climate differs from one area of Earth to another. Some areas have little rain and high temperatures. Other areas have low temperatures and lots of snow. Where this tree grows—on Humphrey Head Point in England—there is constant wind.

- What are the characteristics of different climates?
- What factors affect the climate of a region?
- What is climate and how does it impact life on Earth?

Get Ready to Read

What do you think?
Before you read, decide if you agree or disagree with each of these statements. As you read this chapter, see if you change your mind about any of the statements.

1. Locations at the center of large continents usually have the same climate as locations along the coast.
2. Latitude does not affect climate.
3. Climate on Earth today is the same as it has been in the past.
4. Climate change occurs in short-term cycles.
5. Human activities can impact climate.
6. You can help reduce the amount of greenhouse gases released into the atmosphere.

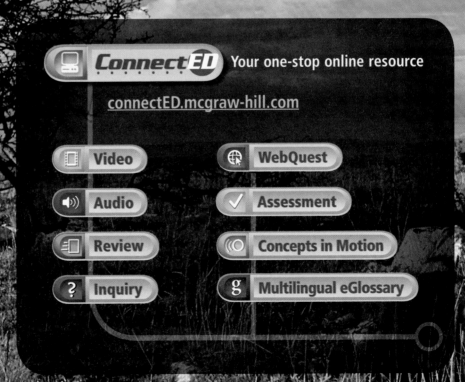

ConnectED — Your one-stop online resource

connectED.mcgraw-hill.com

- Video
- WebQuest
- Audio
- Assessment
- Review
- Concepts in Motion
- Inquiry
- Multilingual eGlossary

Lesson 1

Climates of Earth

Reading Guide

Key Concepts 🔑
ESSENTIAL QUESTIONS

- What is climate?
- Why is one climate different from another?
- How are climates classified?

Vocabulary
climate p. 487
rain shadow p. 489
specific heat p. 489
microclimate p. 491

 Multilingual eGlossary

Inquiry What makes a desert a desert?

How much precipitation do deserts get? Are deserts always hot? What types of plants grow in the desert? Scientists look at the answers to all these questions to determine if an area is a desert.

Launch Lab

20 minutes

How do climates compare?

Climate describes long-term weather patterns for an area. Temperature and precipitation are two factors that help determine climate.

1. Read and complete a lab safety form.
2. Select a location on a **globe.**
3. Research the average monthly temperatures and levels of precipitation for this location.
4. Record your data in a chart like the one shown here in your Science Journal.

Omsk, Russia
73.5° E, 55° N

Month	Average Monthly Temperature	Average Monthly Level of Precipitation
January	−14°C	13 mm
February	−12°C	9 mm
March	−5°C	9 mm
April	8°C	18 mm
May	18°C	31 mm
June	24°C	52 mm
July	25°C	61 mm
August	22°C	50 mm
September	17°C	32 mm
October	7°C	26 mm
November	−4°C	19 mm
December	−12°C	15 mm

Think About This

1. Describe the climate of your selected location in terms of temperature and precipitation.
2. Compare your data to Omsk, Russia. How do the climates differ?
3. **Key Concept** Mountains, oceans, and latitude can affect climates. Do any of these factors account for the differences you observed? Explain.

What is climate?

You probably already know that the term *weather* describes the atmospheric conditions and short term changes of a certain place at a certain time. The weather changes from day to day in many places on Earth. Other places on Earth have more constant weather. For example, temperatures in Antarctica rarely are above 0°C, even in the summer. Areas in Africa's Sahara, shown in the photo on the previous page, have temperatures above 20°C year-round.

Climate *is the long-term average weather conditions that occur in a particular region.* A region's climate depends on average temperature and precipitation, as well as how these variables change throughout the year.

What affects climate?

Several factors determine a region's climate. The latitude of a location affects climate. For example, areas close to the equator have the warmest climates. Large bodies of water, including lakes and oceans, also influence the climate of a region. Along coastlines, weather is more constant throughout the year. Hot summers and cold winters typically happen in the center of continents. The altitude of an area affects climate. Mountainous areas are often rainy or snowy. Buildings and concrete, which retain solar energy, cause temperatures to be higher in urban areas. This creates a special climate in a small area.

Key Concept Check What is climate?

Lesson 1
487
EXPLORE

Figure 1 Latitudes near the poles receive less solar energy and have lower average temperatures.

Latitude

Recall that, starting at the equator, latitude increases from 0° to 90° as you move toward the North Pole or the South Pole. The amount of solar energy per unit of Earth's surface area depends on latitude. **Figure 1** shows that locations close to the equator receive more solar energy per unit of surface area annually than locations located farther north or south. This is due mainly to the fact that Earth's curved surface causes the angle of the Sun's rays to spread out over a larger area. Locations near the equator also tend to have warmer climates than locations at higher latitudes. Polar regions are colder because annually they receive less solar energy per unit of surface area. In the middle latitudes, between 30° and 60°, summers are generally hot and winters are usually cold.

Altitude

Climate is also influenced by altitude. Recall that temperature decreases as altitude increases in the troposphere. So, as you climb a tall mountain you might experience the same cold, snowy climate that is near the poles. **Figure 2** shows the difference in average temperatures between two cities in Colorado at different altitudes.

Altitude and Climate

Figure 2 As altitude increases, temperature decreases.

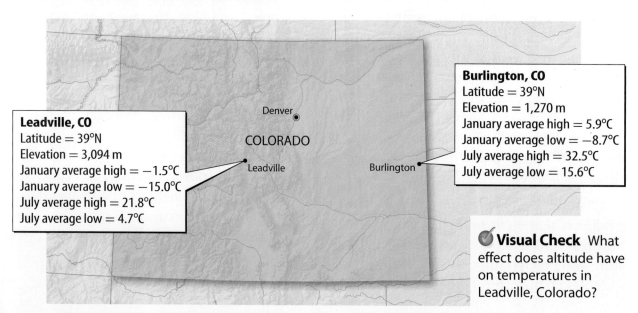

Leadville, CO
Latitude = 39°N
Elevation = 3,094 m
January average high = −1.5°C
January average low = −15.0°C
July average high = 21.8°C
July average low = 4.7°C

Burlington, CO
Latitude = 39°N
Elevation = 1,270 m
January average high = 5.9°C
January average low = −8.7°C
July average high = 32.5°C
July average low = 15.6°C

Visual Check What effect does altitude have on temperatures in Leadville, Colorado?

Rain Shadow 🔑

1. Prevailing winds carry moist, warm air over Earth's surface.

2. As the air approaches mountains, it rises and cools. Water vapor in the air condenses. Precipitation falls as rain or snow on the upwind slope of the mountains.

3. The now-dry air passes over the mountains. As it sinks, it warms.

4. Dry weather exists on the downwind slope of the mountains.

Figure 3 Rain shadows form on the downwind slope of a mountain.

✓ **Visual Check** Why don't rain shadows form on the upwind slope of mountains?

Rain Shadows

Mountains influence climate because they are barriers to prevailing winds. This leads to unique **precipitation** patterns called rain shadows. *An area of low rainfall on the downwind slope of a mountain is called a* **rain shadow**, as shown in **Figure 3**. Different amounts of precipitation on either side of a mountain range influence the types of vegetation that grow. Abundant amounts of vegetation grow on the side of the mountain exposed to the precipitation. The amount of vegetation on the downwind slope is sparse due to the dry weather.

REVIEW VOCABULARY

precipitation
water, in liquid or solid form, that falls from the atmosphere

Large Bodies of Water

On a sunny day at the beach, why does the sand feel warmer than the water? It is because water has a high specific heat. **Specific heat** *is the amount (joules) of thermal energy needed to raise the temperature of 1 kg of a material by 1°C*. The specific heat of water is about six times higher than the specific heat of sand. This means the ocean water would have to absorb six times as much thermal energy to be the same temperature as the sand.

The high specific heat of water causes the climates along coastlines to remain more constant than those in the middle of a continent. For example, the West Coast of the United States has moderate temperatures year-round.

Ocean currents can also modify climate. The Gulf Stream is a warm current flowing northward along the coast of eastern North America. It brings warmer temperatures to portions of the East Coast of the United States and parts of Europe.

✓ **Reading Check** How do large bodies of water influence climate?

Lesson 1
EXPLAIN

World Climates

Figure 4 The map shows a modified version of Köppen's climate classification system.

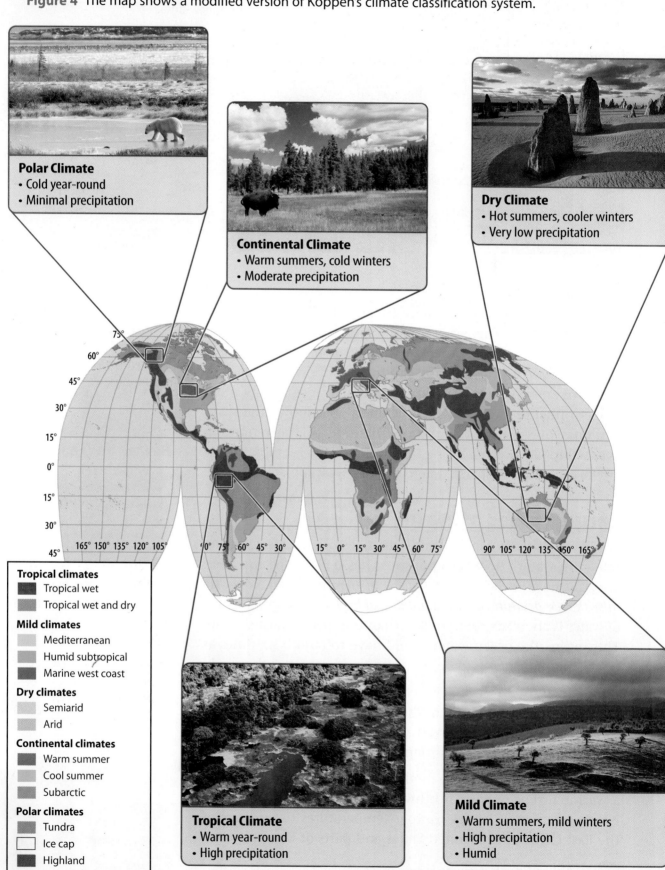

Classifying Climates

What is the climate of any particular region on Earth? This can be a difficult question to answer because many factors affect climate. In 1918 German scientist Wladimir Köppen (vlah DEE mihr • KAWP pehn) developed a system for classifying the world's many climates. Köppen classified a region's climate by studying its temperature, precipitation, and native vegetation. Native vegetation is often limited to particular climate conditions. For example, you would not expect to find a warm-desert cactus growing in the cold, snowy arctic. Köppen identified five climate types. A modified version of Köppen's classification system is shown in **Figure 4**.

 Key Concept Check How are climates classified?

Microclimates

Roads and buildings in cities have more concrete than surrounding rural areas. The concrete absorbs solar radiation, causing warmer temperatures than in the surrounding countryside. The result is a common microclimate called the urban heat island, as shown in **Figure 5**. A **microclimate** *is a localized climate that is different from the climate of the larger area surrounding it.* Other examples of microclimates include forests, which are often cooler and less windy than the surrounding countryside, and hilltops, which are windier than nearby lower land.

 Key Concept Check Why is one climate different from another?

FOLDABLES

Use three sheets of notebook paper to make a layered book. Label it as shown. Use it to organize your notes on the factors that determine a region's climate.

WORD ORIGIN

microclimate
from Greek *mikros*, means "small"; and *klima*, means "region, zone"

Microclimate

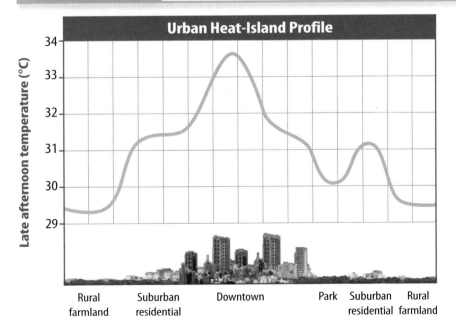

Figure 5 The temperature is often warmer in urban areas when compared to temperatures in the surrounding countryside.

Visual Check What is the temperature difference between downtown and rural farmland?

Lesson 1
EXPLAIN

Figure 6 Camels are adapted to dry climates and can survive up to three weeks without drinking water.

How Climate Affects Living Organisms

Organisms have adaptations for the climates where they live. For example, polar bears have thick fur and a layer of fat that helps keep them warm in the Arctic. Many animals that live in deserts, such as the camels in **Figure 6,** have adaptations for surviving in hot, dry conditions. Some desert plants have extensive shallow root systems that collect rainwater. Deciduous trees, found in continental climates, lose their leaves during the winter, which reduces water loss when soils are frozen.

Climate also influences humans in many ways. Average temperature and rainfall in a location help determine the type of crops humans grow there. Thousands of orange trees grow in Florida, where the climate is mild. Wisconsin's continental climate is ideal for growing cranberries.

Climate also influences the way humans design buildings. In polar climates, the soil is frozen year-round—a condition called permafrost. Humans build houses and other buildings in these climates on stilts. This is done so that thermal energy from the building does not melt the permafrost.

Reading Check How are organisms adapted to different climates?

Inquiry MiniLab

40 minutes

Where are microclimates found?

Microclimates differ from climates in the larger region around them. In this lab, you will identify a microclimate.

1. Read and complete a lab safety form.
2. Select two areas near your school. One area should be in an open location. The other area should be near the school building.
3. Make a data table like the one at the right in your Science Journal.
4. Measure and record data at the first area. Find wind direction using a **wind sock,** temperature using a **thermometer,** and relative humidity using a **psychrometer** and a **relative humidity chart.**
5. Repeat step 4 at the second area.

	Sidewalk	Soccer Fields
Temperature		
Wind direction		
Relative humidity		

Analyze and Conclude

1. **Graph Data** Make a bar graph showing the temperature and relative humidity at both sites.
2. **Use** the data in your table to compare wind direction.
3. **Interpret Data** How did weather conditions at the two sites differ? What might account for these differences?
4. **Key Concept** How might you decide which site is a microclimate? Explain.

Lesson 1 Review

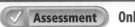 Assessment   Online Quiz

Visual Summary

Climate is influenced by several factors including latitude, altitude, and an area's location relative to a large body of water or mountains.

Rain shadows occur on the downwind slope of mountains.

Microclimates can occur in urban areas, forests, and hilltops.

FOLDABLES

Use your lesson Foldable to review the lesson. Save your Foldable for the project at the end of the chapter.

What do you think NOW?

You first read the statements below at the beginning of the chapter.

1. Locations at the center of large continents usually have the same climate as locations along the coast.
2. Latitude does not affect climate.

Did you change your mind about whether you agree or disagree with the statements? Rewrite any false statements to make them true.

Use Vocabulary

1. The amount of thermal energy needed to raise the temperature of 1 kg of a material by 1°C is called _____.

2. **Distinguish** between climate and microclimate.

3. **Use the term** *rain shadow* in a sentence.

Understand Key Concepts

4. How are climates classified?
   A. by cold- and warm-water ocean currents
   B. by latitude and longitude
   C. by measurements of temperature and humidity
   D. by temperature, precipitation, and vegetation

5. **Describe** the climate of an island in the tropical Pacific Ocean.

6. **Compare** the climates on either side of a large mountain range.

7. **Distinguish** between weather and climate.

Interpret Graphics

8. **Summarize** Copy and fill in the graphic organizer below to summarize information about the different types of climate worldwide.

Climate Type	Description
Tropical	
Dry	
Mild	
Continental	
Polar	

Critical Thinking

9. **Distinguish** between the climates of a coastal location and a location in the center of a large continent.

10. **Infer** how you might snow ski on the island of Hawaii.

Lesson 1 • 493
EVALUATE

Inquiry Skill Practice — Infer

40 minutes

Can reflection of the Sun's rays change the climate?

Materials

bowl

polyester film

transparent tape

stopwatch

light source

thermometer

Safety

Albedo is the term used to refer to the percent of solar energy that is reflected back into space. Clouds, for example, reflect about 50 percent of the solar energy they receive, whereas dark surfaces on Earth might reflect as little as 5 percent. Snow has a very high albedo and reflects 75 to 90 percent of the solar energy it receives. The differences in how much solar energy is reflected back into the atmosphere from different regions of Earth can cause differences in climate. Also, changes in albedo can affect the climate of that region.

Learn It

When an observation cannot be made directly, a simulation can be used to draw reasonable conclusions. This strategy is known as **inferring**. Simulating natural occurrences on a small scale can provide indirect observations so realistic outcomes can be inferred.

Try It

1. Read and complete a lab safety form.
2. Make a data table for recording temperatures in your Science Journal.
3. Cover the bottom of a bowl with a sheet of polyester film. Place a thermometer on top of the sheet. Record the temperature in the bottom of the bowl.
4. Put the bowl under the light source and set the timer for 5 minutes. After 5 minutes, record the temperature. Remove the thermometer and allow it to return to its original temperature. Repeat two more times.
5. Repeat the experiment, but this time tape the sheet of polyester film over the top of the bowl and the thermometer.

Apply It

6. **Analyze** the data you collected. What difference did you find when the polyester film covered the bowl?
7. **Conclude** What can you conclude about the Sun's rays reaching the bottom of the bowl when it was covered by the polyester film?
8. **Infer** what happens to the Sun's rays when they reach clouds in the atmosphere. Explain.
9. **Describe** how the high albedo of the ice and snow in the polar regions contribute to the climate there.
10. **Key Concept** If a region of Earth were to be covered most of the time by smog or clouds, would the climate of that region change? Explain your answer.

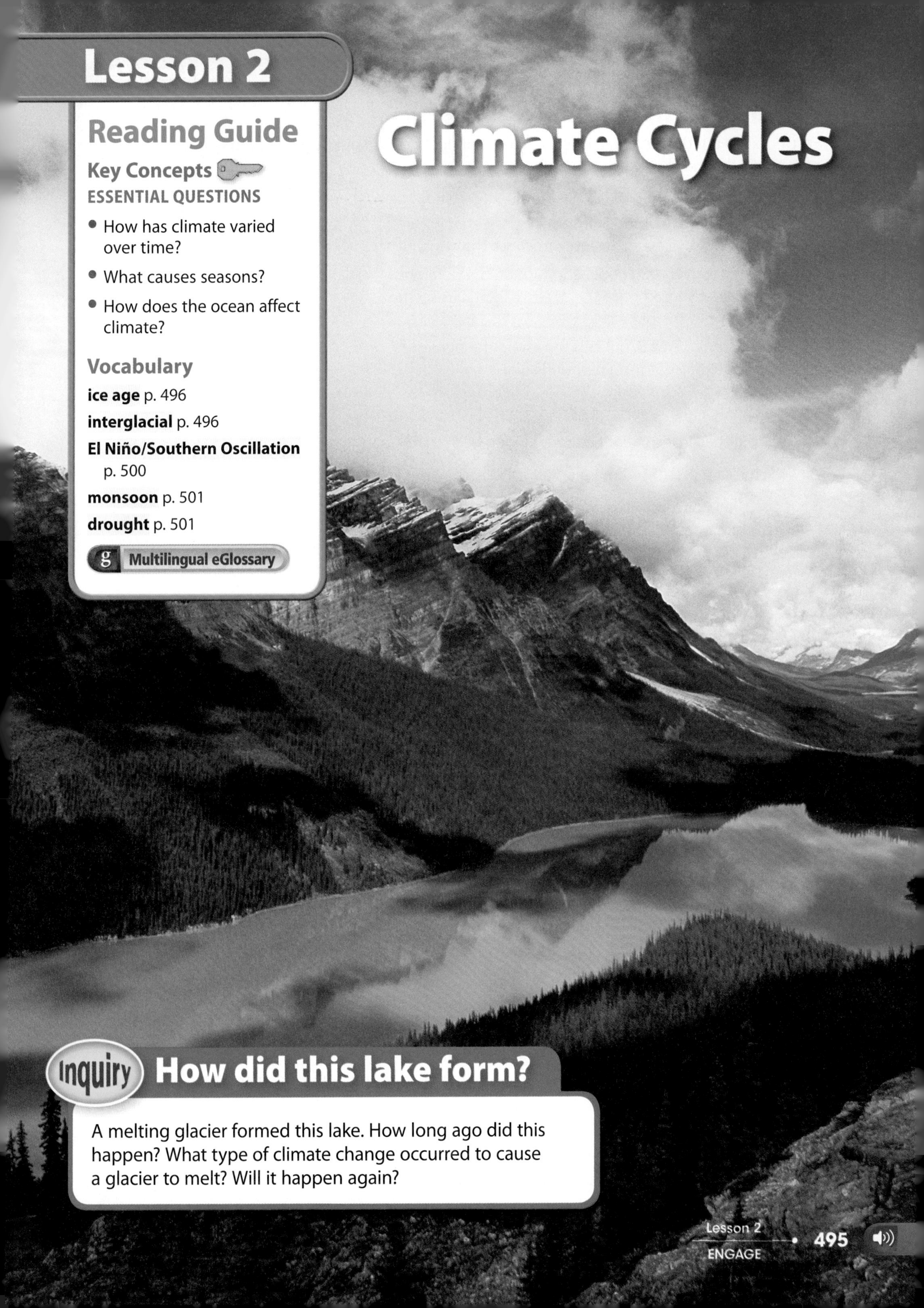

Lesson 2

Climate Cycles

Reading Guide

Key Concepts
ESSENTIAL QUESTIONS

- How has climate varied over time?
- What causes seasons?
- How does the ocean affect climate?

Vocabulary

ice age p. 496
interglacial p. 496
El Niño/Southern Oscillation p. 500
monsoon p. 501
drought p. 501

Multilingual eGlossary

Inquiry How did this lake form?

A melting glacier formed this lake. How long ago did this happen? What type of climate change occurred to cause a glacier to melt? Will it happen again?

Launch Lab

20 minutes

How does Earth's tilted axis affect climate?
Earth's axis is tilted at an angle of 23.5°. This tilt influences climate by affecting the amount of sunlight that reaches Earth's surface.

1. Read and complete a lab safety form.
2. Hold a **penlight** about 25 cm above a sheet of paper at a 90° angle. Use a **protractor** to check the angle.
3. Turn off the overhead lights and turn on the penlight. Your partner should trace the circle of light cast by the penlight onto the paper.
4. Repeat steps 2 and 3, but this time hold the penlight at an angle of 23.5° from perpendicular.

Think About This
1. How did the circles of light change during each trial?
2. Which trial represented the tilt of Earth's axis?
3. **Key Concept** How might changes in the tilt of Earth's axis affect climate? Explain.

Figure 7 Scientists study the different layers in an ice core to learn more about climate changes in the past.

Long-Term Cycles

Weather and climate have many cycles. In most areas on Earth, temperatures increase during the day and decrease at night. Each year, the air is warmer during summer and colder during winter. But climate also changes in cycles that take much longer than a lifetime to complete.

Much of our knowledge about past climates comes from natural records of climate. Scientists study ice cores, shown in **Figure 7,** drilled from ice layers in glaciers and ice sheets. Fossilized pollen, ocean sediments, and the growth rings of trees also are used to gain information about climate changes in the past. Scientists use the information to compare present-day climates to those that occurred many thousands of years ago.

Reading Check How do scientists find information about past climates on Earth?

Ice Ages and Interglacials
Earth has experienced many major atmospheric and climate changes in its history. **Ice ages** are cold periods lasting from hundreds to millions of years when glaciers cover much of Earth. Glaciers and ice sheets advance during cold periods and retreat during **interglacials**—*the warm periods that occur during ice ages or between ice ages.*

Major Ice Ages and Warm Periods

The most recent ice age began about 2 million years ago. The ice sheets reached maximum size about 20,000 years ago. At that time, about half the northern hemisphere was covered by ice. About 10,000 years ago, Earth entered its current interglacial period, called the Holocene Epoch.

Temperatures on Earth have fluctuated during the Holocene. For example, the period between 950 and 1100 was one of the warmest in Europe. The Little Ice Age, which lasted from 1250 to about 1850, was a period of bitterly cold temperatures.

 Key Concept Check How has climate varied over time?

Causes of Long-Term Climate Cycles

As the amount of solar energy reaching Earth changes, Earth's climate also changes. One factor that affects how much energy Earth receives is the shape of its orbit. The shape of Earth's orbit appears to vary between elliptical and circular over the course of about 100,000 years. As shown in **Figure 8,** when Earth's orbit is more circular, Earth averages a greater distance from the Sun. This results in below-average temperatures on Earth.

Another factor that scientists suspect influences climate change on Earth is changes in the tilt of Earth's axis. The tilt of Earth's axis changes in 41,000-year cycles. Changes in the angle of Earth's tilt affect the range of temperatures throughout the year. For example, a decrease in the angle of Earth's tilt, as shown in **Figure 8,** could result in a decrease in temperature differences between summer and winter. Long-term climate cycles are also influenced by the slow movement of Earth's continents, as well as changes in ocean circulation.

> **WORD ORIGIN**
> **interglacial**
> from Latin *inter–*, means "among, between"; and *glacialis*, means "icy, frozen"

Figure 8 This exaggerated image shows how the shape of Earth's orbit varies between elliptical and circular. The angle of the tilt varies from 22° to 24.5° about every 41,000 years. Earth's current tilt is 23.5°.

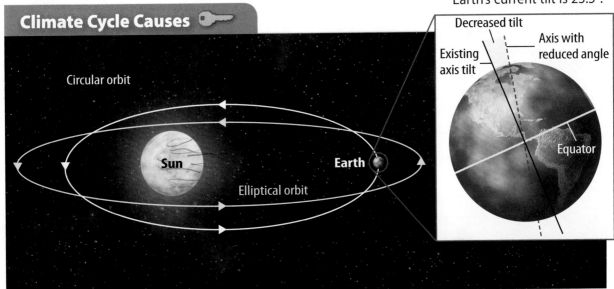

Short-Term Cycles

In addition to its long-term cycles, climate also changes in short-term cycles. Seasonal changes and changes that result from the interaction between the ocean and the atmosphere are some examples of short-term climate change.

Seasons

Changes in the amount of solar energy received at different latitudes during different times of the year give rise to the seasons. Seasonal changes include regular changes in temperature and the number of hours of day and night.

Recall from Lesson 1 that the amount of solar energy per unit of Earth's surface is related to latitude. Another factor that affects the amount of solar energy received by an area is the tilt of Earth's axis. **Figure 9** shows that when the northern hemisphere is tilted toward the Sun, there are more daylight hours than dark hours, and temperatures are warmer. The northern hemisphere receives more direct solar energy and it is summer. At the same time, the southern hemisphere receives less overall solar energy and it is winter there.

Figure 9 shows that the opposite occurs when six months later the northern hemisphere is tilted away from the Sun. Daylight hours are fewer than nighttime hours, and temperatures are colder. Indirect solar energy reaches the northern hemisphere, resulting in winter. The southern hemisphere receives more direct solar energy and it is summer.

Key Concept Check What causes seasons?

FOLDABLES
Make a horizontal three-tab book and label it as shown. Use your book to organize information about short-term climate cycles. Fold the book into thirds and label the outside *Short Term Climate Cycles*.

Figure 9 The solar energy rays reaching a given area of Earth's surface is more intense when tilted toward the Sun.

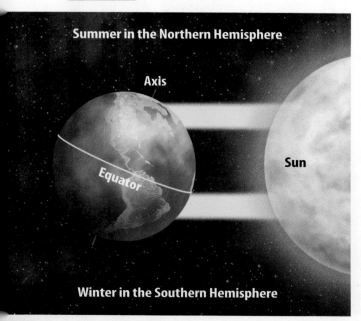
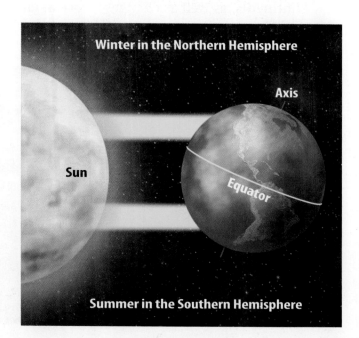

Northern Hemisphere Seasons

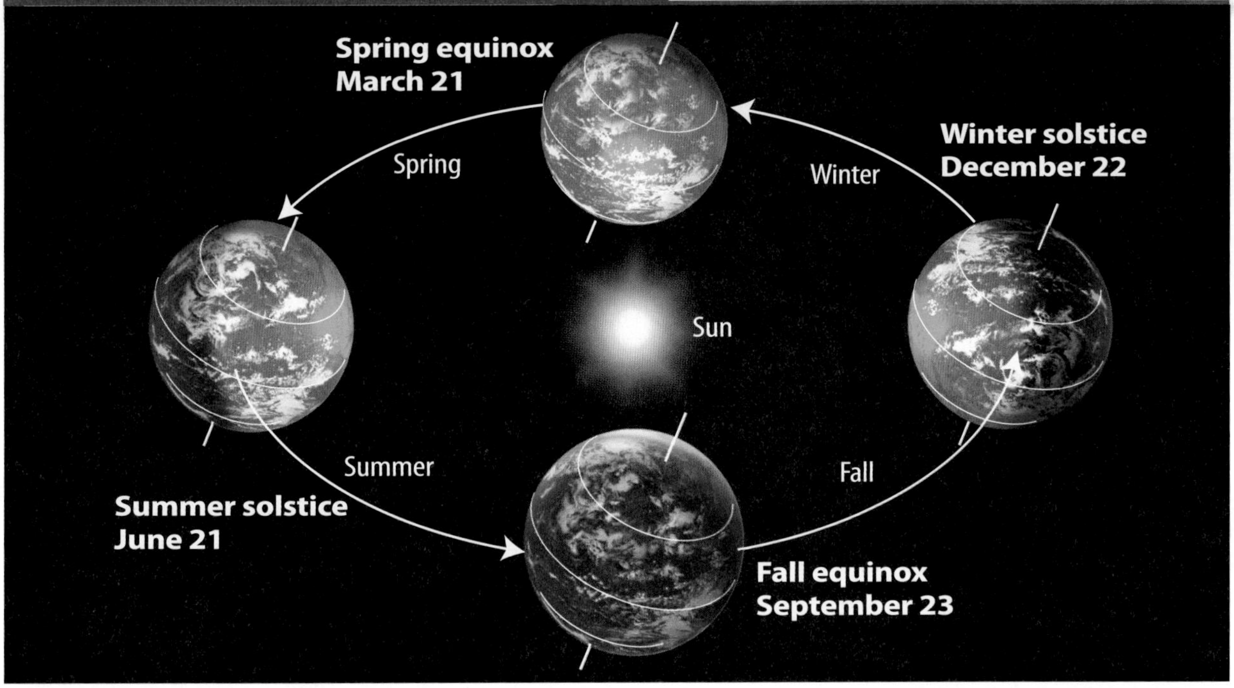

Figure 10 Seasons change as Earth completes its yearly revolution around the Sun.

Visual Check How does the amount of sunlight striking the North Pole change from summer to winter?

Solstices and Equinoxes

Earth revolves around the Sun once about every 365 days. During Earth's **revolution,** there are four days that mark the beginning of each of the seasons. These days are a summer solstice, a fall equinox, a winter solstice, and a spring equinox.

As shown in **Figure 10,** the solstices mark the beginnings of summer and winter. In the northern hemisphere, the summer solstice occurs on June 21 or 22. On this day, the northern hemisphere is tilted toward the Sun. In the southern hemisphere, this day marks the beginning of winter. The winter solstice begins on December 21 or 22 in the northern hemisphere. On this day, the northern hemisphere is tilted away from the Sun. In the southern hemisphere, this day marks the beginning of summer.

Equinoxes, also shown in **Figure 10,** are days when Earth is positioned so that neither the northern hemisphere nor the southern hemisphere is tilted toward or away from the Sun. The equinoxes are the beginning of spring and fall. On equinox days, the number of daylight hours almost equals the number of nighttime hours everywhere on Earth. In the northern hemisphere, the spring equinox occurs on March 21 or 22. This is the beginning of fall in the southern hemisphere. On September 22 or 23, fall begins in the northern hemisphere and spring begins in the southern hemisphere.

Reading Check Compare and contrast solstices and equinoxes.

SCIENCE USE V. COMMON USE

revolution

Science Use the action by a celestial body of going around in an orbit or an elliptical course

Common Use a sudden, radical, or complete change

El Niño

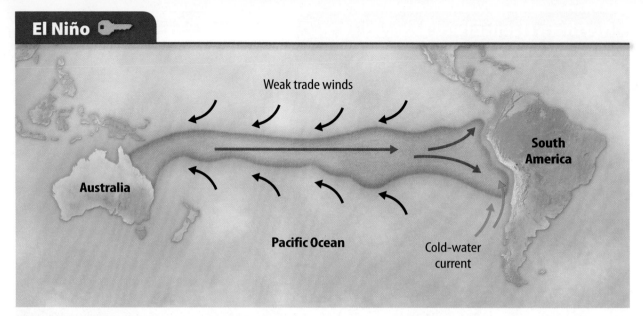

Figure 11 During El Niño, the trade winds weaken and warm water surges toward South America.

Visual Check Where is the warm water during normal conditions?

ACADEMIC VOCABULARY

phenomenon
(noun) an observable fact or event

El Niño and the Southern Oscillation

Close to the equator, the trade winds blow from east to west. These steady winds push warm surface water in the Pacific Ocean away from the western coast of South America. This allows cold water to rush upward from below—a process called upwelling. The air above the cold, upwelling water cools and sinks, creating a high-pressure area. On the other side of the Pacific Ocean, air rises over warm, equatorial waters, creating a low-pressure area. This difference in air pressures across the Pacific Ocean helps keep the trade winds blowing.

As **Figure 11** shows, sometimes the trade winds weaken, reversing the normal pattern of high and low pressures across the Pacific Ocean. Warm water surges back toward South America, preventing cold water from upwelling. This phenomenon, called El Niño, shows the connection between the atmosphere and the ocean. During El Niño, the normally dry, cool western coast of South America warms and receives lots of precipitation. Climate changes can be seen around the world. Droughts occur in areas that are normally wet. The number of violent storms in California and the southern United States increases.

Reading Check How do conditions in the Pacific Ocean differ from normal during El Niño?

The combined ocean and atmospheric cycle that results in weakened trade winds across the Pacific Ocean is called **El Niño/Southern Oscillation,** or ENSO. A complete ENSO cycle occurs every 3–8 years. The North Atlantic Oscillation (NAO) is another cycle that can change the climate for decades at a time. The NAO affects the strength of storms throughout North America and Europe by changing the position of the jet stream.

Monsoons

Another climate cycle involving both the atmosphere and the ocean is a monsoon. A **monsoon** *is a wind circulation pattern that changes direction with the seasons.* Temperature differences between the ocean and the land cause winds, as shown in **Figure 12**. During summer, warm air over land rises and creates low pressure. Cooler, heavier air sinks over the water, creating high pressure. The winds blow from the water toward the land, bringing heavy rainfall. During winter, the pattern reverses and winds blow from the land toward the water.

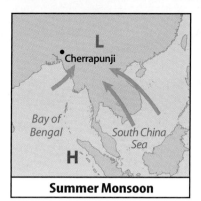
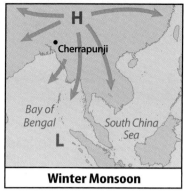

Figure 12 Monsoon winds reverse with the change of seasons.

The world's largest monsoon is found in Asia. Cherrapunji, India, is one of the world's wettest locations—receiving an average of 10 m of monsoon rainfall each year. Precipitation is even greater during El Niño events. A smaller monsoon occurs in southern Arizona. As a result, weather is dry during spring and early summer with thunderstorms occurring more often from July to September.

 Key Concept Check How does the ocean affect climate?

Droughts, Heat Waves, and Cold Waves

A **drought** *is a period with below-average precipitation.* A drought can cause crop damage and water shortages.

Droughts are often accompanied by heat waves—periods of unusually high temperatures. Droughts and heat waves occur when large hot-air masses remain in one place for weeks or months. Cold waves are long periods of unusually cold temperatures. These events occur when a large continental polar air mass stays over a region for days or weeks. Severe weather of these kinds can be the result of climatic changes on Earth or just extremes in the average weather of a climate.

Inquiry MiniLab
20 minutes

How do climates vary?
Unlike El Niño, La Niña is associated with cold ocean temperatures in the Pacific Ocean.

1. As the map shows, average temperatures change during a La Niña winter.
2. The color key shows the range of temperature variation from normal.
3. Find a location on the map. How much did temperatures during La Niña depart from average temperatures?

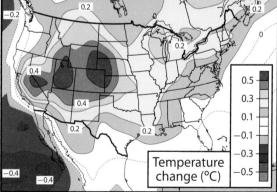

Analyze and Conclude

1. **Recognize Cause and Effect** Did La Niña affect the climate in your chosen area?
2. **Key Concept** Describe any patterns you see. How did La Niña affect climate in your chosen area? Use data from the map to support your answer.

Lesson 2
EXPLAIN

Lesson 2 Review

Assessment | Online Quiz

Visual Summary

Scientists learn about past climates by studying natural records of climate, such as ice cores, fossilized pollen, and growth rings of trees.

Long-term climate changes, such as ice ages and interglacials, can be caused by changes in the shape of Earth's orbit and the tilt of its axis.

Short-term climate changes include seasons, El Niño/Southern Oscillation, and monsoons.

FOLDABLES

Use your lesson Foldable to review the lesson. Save your Foldable for the project at the end of the chapter.

What do you think NOW?

You first read the statements below at the beginning of the chapter.

3. Climate on Earth today is the same as it has been in the past.

4. Climate change occurs in short-term cycles.

Did you change your mind about whether you agree or disagree with the statements? Rewrite any false statements to make them true.

Use Vocabulary

1. **Distinguish** an ice age from an interglacial.
2. A(n) _____ is a period of unusually high temperatures.
3. **Define** *drought* in your own words.

Understand Key Concepts

4. What happens during El Niño/Southern Oscillation?
   A. An interglacial climate shift occurs.
   B. The Pacific pressure pattern reverses.
   C. The tilt of Earth's axis changes.
   D. The trade winds stop blowing.
5. **Identify** causes of long-term climate change.
6. **Describe** how upwelling can affect climate.

Interpret Graphics

7. **Sequence** Copy and fill in the graphic organizer below to describe the sequence of events during El Niño/Southern Oscillation.

Critical Thinking

8. **Assess** the possibility that Earth will soon enter another ice age.
9. **Evaluate** the relationship between heat waves and drought.
10. **Identify** and explain the climate cycle shown below. Illustrate how conditions change during the summer.

Winter Monsoon

502 Chapter 14 EVALUATE

AMERICAN MUSEUM OF NATURAL HISTORY

Frozen in Time

CAREERS in SCIENCE

Looking for clues to past climates, Lonnie Thompson races against the clock to collect ancient ice from melting glaciers.

Earth's climate is changing. To understand why, scientists investigate how climates have changed throughout Earth's history by looking at ancient ice that contains clues from past climates. Scientists collected these ice samples only from glaciers at the North Pole and the South Pole. Then, in the 1970s, geologist Lonnie Thompson began collecting ice from a new location—the tropics.

Thompson, a geologist from the Ohio State University, and his team scale glaciers atop mountains in tropical regions. On the Quelccaya ice cap in Peru, they collect ice cores—columns of ice layers that built up over hundreds to thousands of years. Each layer is a capsule of a past climate, holding dust, chemicals, and gas that were trapped in the ice and snow during that period.

To collect ice cores, they drill hundreds of feet into the ice. The deeper they drill, the further back in time they go. One core is nearly 12,000 years old!

Collecting ice cores is not easy. The team hauls heavy equipment up rocky slopes in dangerous conditions—icy windstorms, thin air, and avalanche threats. Thompson's greatest challenge is the warming climate. The Quelccaya ice cap is melting. It has shrunk by 30 percent since Thompson's first visit in 1974. It's a race against time to collect ice cores before the ice disappears. When the ice is gone, so are the secrets it holds about climate change.

◀ Thompson has led expeditions to 15 countries and Antarctica.

Thousands of ice core samples are stored in deep freeze at Thompson's lab. One core from Antarctica is over 700,000 years old, which is well before the existence of humans. ▶

Secrets in the Ice

In the lab, Thompson and his team analyze the ice cores to determine

- **Age of ice:** Every year, snow accumulations form a new layer. Layers help scientists date the ice and specific climate events.

- **Precipitation:** Each layer's thickness and composition help scientists determine the amount of snowfall that year.

- **Atmosphere:** As snow turns to ice, it traps air bubbles, providing samples of the Earth's atmosphere. Scientists can measure the trace gases from past climates.

- **Climate events:** The concentration of dust particles helps scientists determine periods of increased wind, volcanic activity, dust storms, and fires.

It's Your Turn

WRITE AN INTRODUCTION Imagine Lonnie Thompson is giving a speech at your school. You have been chosen to introduce him. Write an introduction highlighting his work and achievements.

Lesson 2 EXTEND

Lesson 3

Reading Guide

Key Concepts
ESSENTIAL QUESTIONS

- How can human activities affect climate?
- How are predictions for future climate change made?

Vocabulary
global warming p. 506
greenhouse gas p. 506
deforestation p. 507
global climate model p. 509

 Multilingual eGlossary

Video BrainPOP®

Recent Climate Change

Inquiry Will Tuvalu sink or swim?

This small island sits in the middle of the Pacific Ocean. What might happen to this island if the sea level rose? What type of climate change might cause sea level to rise?

Launch Lab

20 minutes

What changes climates?

Natural events such as volcanic eruptions spew dust and gas into the atmosphere. These events can cause climate change.

1. Read and complete a lab safety form.
2. Place a **thermometer** on a sheet of **paper.**
3. Hold a **flashlight** 10 cm above the paper. Shine the light on the thermometer bulb for 5 minutes. Observe the light intensity. Record the temperature in your Science Journal.
4. Use a **rubber band** to secure 3–4 layers of **cheesecloth or gauze** over the bulb end of the flashlight. Repeat step 3.

Think About This

1. Describe the effect of the cheesecloth on the flashlight in terms of brightness and temperature.

2. **Key Concept** Would a volcanic eruption cause temperatures to increase or decrease? Explain.

Regional and Global Climate Change

Average temperatures on Earth have been increasing for the past 100 years. As the graph in **Figure 13** shows, the warming has not been steady. Globally, average temperatures were fairly steady from 1880 to 1900. From 1900 to 1945, they increased by about 0.5°C. A cooling period followed, ending in 1975. Since then, average temperatures have steadily increased. The greatest warming has been in the northern hemisphere. However, temperatures have been steady in some areas of the southern hemisphere. Parts of Antarctica have cooled.

Reading Check How have temperatures changed over the last 100 years?

FOLDABLES

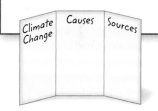

Make a tri-fold book from a sheet of paper. Label it as shown. Use it to organize your notes about climate change and the possible causes.

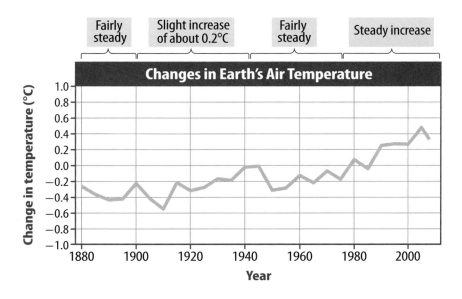

Figure 13 Temperature change has not been constant throughout the past 100 years.

Visual Check What 20-year period has seen the most change?

Lesson 3

EXPLORE

505

Human Impact on Climate Change

The rise in Earth's average surface temperature during the past 100 years is often referred to as **global warming.** Scientists have been studying this change and the possible causes of it. In 2007, the Intergovernmental Panel on Climate Change (IPCC), an international organization created to study global warming, concluded that most of this temperature increase is due to human activities. These activities include the release of increasing amounts of greenhouse gases into the atmosphere through burning fossil fuels and large-scale cutting and burning of forests. Although many scientists agree with the IPCC, some scientists propose that global warming is due to natural climate cycles.

WORD ORIGIN

deforestation
from Latin *de-*, means "down from, concerning"; and *forestum silvam*, means "the outside woods"

Greenhouse Gases

Gases in the atmosphere that absorb Earth's outgoing infrared radiation are **greenhouse gases.** Greenhouse gases help keep temperatures on Earth warm enough for living things to survive. Recall that this phenomenon is referred to as the greenhouse effect. Without greenhouse gases, the average temperature on Earth would be much colder, about −18°C. Carbon dioxide (CO_2), methane, and water vapor are all greenhouse gases.

Study the graph in **Figure 14.** What has happened to the levels of CO_2 in the atmosphere over the last 120 years? Levels of CO_2 have been increasing. Higher levels of greenhouse gases create a greater greenhouse effect. Most scientists suggest that global warming is due to the greater greenhouse effect. What are some sources of the excess CO_2?

 Reading Check How do greenhouse gases affect temperatures on Earth?

Climate Change

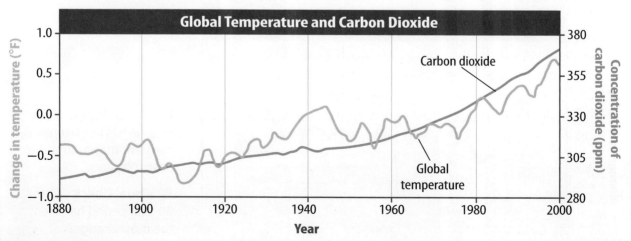

Figure 14 Over the recent past, globally averaged temperatures and carbon dioxide concentration in the atmosphere have both increased.

Human-Caused Sources Carbon dioxide enters the atmosphere when fossil fuels, such as coal, oil, and natural gas, burn. Burning fossil fuels releases energy that provides electricity, heats homes and buildings, and powers automobiles.

Deforestation *is the large-scale cutting and/or burning of forests.* Forest land is often cleared for agricultural and development purposes. Deforestation, shown in **Figure 15**, affects global climate by increasing carbon dioxide in the atmosphere in two ways. Living trees remove carbon dioxide from the air during photosynthesis. Cut trees, however, do not. Sometimes cut trees are burned to clear a field, adding carbon dioxide to the atmosphere as the trees burn. According to the Food and Agriculture Organization of the United Nations, deforestation makes up about 25 percent of the carbon dioxide released from human activities.

Natural Sources Carbon dioxide occurs naturally in the atmosphere. Its sources include volcanic eruptions and forest fires. Cellular respiration in organisms contributes additional CO_2.

Aerosols

The burning of fossil fuels releases more than just greenhouse gases into the atmosphere. Aerosols, tiny liquid or solid particles, are also released. Most aerosols reflect sunlight back into space. This prevents some of the Sun's energy from reaching Earth, potentially cooling the climate over time.

Aerosols also cool the climate in another way. When clouds form in areas with large amounts of aerosols, the cloud droplets are smaller. Clouds with small droplets, as shown in **Figure 16,** reflect more sunlight than clouds with larger droplets. By preventing sunlight from reaching Earth's surface, small-droplet clouds help cool the climate.

Key Concept Check How can human activities affect climate?

▲ **Figure 15** When forests are cut down, trees can no longer use carbon dioxide from the atmosphere. In addition, any wood left rots and releases more carbon dioxide into the atmosphere.

Figure 16 Clouds made up of small droplets reflect more sunlight than clouds made up of larger droplets. ▼

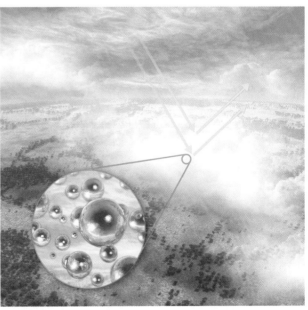

Math Skills

Use Percents
If Earth's population increases from 6 billion to 9 billion, what percent is this increase?

1. Subtract the initial value from the final value:

   9 billion − 6 billion = 3 billion

2. Divide the difference by the starting value:

   $\frac{3 \text{ billion}}{6 \text{ billion}} = 0.50$

3. Multiply by 100 and add a % sign: $0.50 \times 100 = 50\%$

Practice
If a climate's mean temperature changes from 18.2°C to 18.6°C, what is the percentage of increase?

 Review
- Math Practice
- Personal Tutor

Climate and Society

A changing climate can present serious problems for society. Heat waves and droughts can cause food and water shortages. Excessive rainfall can cause flooding and mudslides. However, climate change can also benefit society. Warmer temperatures can mean longer growing seasons. Farmers can grow crops in areas that were previously too cold. Governments throughout the world are responding to the problems and opportunities created by climate change.

Environmental Impacts of Climate Change

Recall that ENSO cycles can change the amount of precipitation in some areas. Warmer ocean surface temperatures can cause more water to evaporate from the ocean surface. The increased water vapor in the atmosphere can result in heavy rainfall and frequent storms in North and South America. Increased precipitation in these areas can lead to decreased precipitation in other areas, such as parts of southern Africa, the Mediterranean, and southern Asia.

Increasing temperatures can also impact the environment in other ways. Melting glaciers and polar ice sheets can cause the sea level to rise. Ecosystems can be disrupted as coastal areas flood. Coastal flooding is a serious concern for the one billion people living in low-lying areas on Earth.

Extreme weather events are also becoming more common. What effect will heat waves, droughts, and heavy rainfall have on infectious disease, existing plants and animals, and other systems of nature? Will increased CO_2 levels work similarly?

The annual thawing of frozen ground has caused the building shown in **Figure 17** to slowly sink as the ground becomes soft and muddy. Permanently higher temperatures would create similar events worldwide. This and other ecosystem changes can affect migration patterns of insects, birds, fish, and mammals.

Figure 17 Buildings in the Arctic that were built on frozen soil are now being damaged by the constant freezing and thawing of the soil.

Predicting Climate Change

Weather forecasts help people make daily choices about their clothing and activities. In a similar way, climate forecasts help governments decide how to respond to future climate changes.

A **global climate model**, *or GCM, is a set of complex equations used to predict future climates.* GCMs are similar to models used to forecast the weather. GCMs and weather forecast models are different. GCMs make long-term, global predictions, but weather forecasts are short-term and can be only regional predictions. GCMs combine mathematics and physics to predict temperature, amount of precipitation, wind speeds, and other characteristics of climate. Powerful supercomputers solve mathematical equations and the results are displayed as maps. GCMs include the effects of greenhouse gases and oceans in their calculations. In order to test climate models, past records of climate change can and have been used.

✓ **Reading Check** What is a GCM?

One drawback of GCMs is that the forecasts and predictions cannot be immediately compared to real data. A weather forecast model can be analyzed by comparing its predictions with meteorological measurements made the next day. GCMs predict climate conditions for several decades in the future. For this reason, it is difficult to evaluate the accuracy of climate models.

Most GCMs predict further global warming as a result of greenhouse gas emissions. By the year 2100, temperatures are expected to rise by between 1°C and 4°C. The polar regions are expected to warm more than the tropics. Summer arctic sea ice is expected to completely disappear by the end of the twenty-first century. Global warming and sea-level rise are predicted to continue for several centuries.

 Key Concept Check How are predictions for future climate change made?

Inquiry MiniLab 30 minutes

How much CO_2 do vehicles emit?

Much of the carbon dioxide emitted into the atmosphere by households comes from gasoline-powered vehicles. Different vehicles emit different amounts of CO_2.

1. To calculate the amount of CO_2 given off by a vehicle, you must know how many miles per gallon of gasoline the vehicle gets. This information is shown in the chart below.

2. Assume that each vehicle is driven about 15,000 miles annually. Calculate how many gallons each vehicle uses per year. Record your data in your Science Journal in a chart like the one below.

3. One gallon of gasoline emits about 20 lbs of CO_2. Calculate and record how many pounds of CO_2 are emitted by each vehicle annually.

	Estimated MPG	Gallons of Gas Used Annually	Amount of CO_2 Emitted Annually (lbs)
SUV	15		
Hybrid	45		
Compact car	25		

Analyze and Conclude

1. **Compare and contrast** the amount of CO_2 emitted by each vehicle.

2. 🔑 **Key Concept** Write a letter to a person who is planning to buy a vehicle. Explain which vehicle would have the least impact on global warming and why.

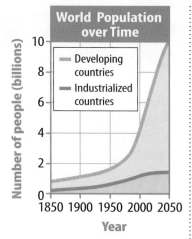

▲ **Figure 18** Earth's population is predicted to increase to more than 9 billion people by 2050.

Human Population

In 2000, more than 6 billion people inhabited Earth. As shown in **Figure 18,** Earth's population is expected to increase to 9 billion by the year 2050. What effects will a 50-percent increase in population have on Earth's atmosphere?

It is predicted that by the year 2030, two of every three people on Earth will live in urban areas. Many of these areas will be in developing countries in Africa and Asia. Large areas of forests are already being cleared to make room for expanding cities. Significant amounts of greenhouse gases and other air pollutants will be added to the atmosphere.

 Reading Check How could an increase in human population affect climate change?

Ways to Reduce Greenhouse Gases

People have many options for reducing levels of pollution and greenhouse gases. One way is to develop alternative sources of energy that do not release carbon dioxide into the atmosphere, such as solar energy or wind energy. Automobile emissions can be reduced by as much as 35 percent by using hybrid vehicles. Hybrid vehicles use an electric motor part of the time, which reduces fuel use.

Emissions can be further reduced by green building. Green building is the practice of creating energy-efficient buildings, such as the one shown in **Figure 19.** People can also help remove carbon dioxide from the atmosphere by planting trees in deforested areas.

You can also help control greenhouse gases and pollution by conserving fuel and recycling. Turning off lights and electronic equipment when you are not using them reduces the amount of electricity you use. Recycling metal, paper, plastic, and glass reduces the amount of fuel required to manufacture these materials.

Figure 19 Solar heating, natural lighting, and water recycling are some of the technologies used in green buildings. ▶

Lesson 3 Review

Assessment — Online Quiz
Inquiry — Virtual Lab

Visual Summary

Many scientists suggest that global warming is due to increased levels of greenhouse gases in atmosphere.

Human activities, such as deforestation and burning fossil fuels, can contribute to global warming.

Ways to reduce greenhouse gas emissions include using solar and wind energy, and creating energy-efficient buildings.

FOLDABLES

Use your lesson Foldable to review the lesson. Save your Foldable for the project at the end of the chapter.

What do you think NOW?

You first read the statements below at the beginning of the chapter.

5. Human activities can impact climate.

6. You can help reduce the amount of greenhouse gases released into the atmosphere.

Did you change your mind about whether you agree or disagree with the statements? Rewrite any false statements to make them true.

Use Vocabulary

1. **Define** *global warming* in your own words.

2. A set of complex equations used to predict future climates is called _____.

3. **Use the term** *deforestation* in a sentence.

Understand Key Concepts

4. Which human activity can have a cooling effect on climate?
   A. release of aerosols
   B. global climate models
   C. greenhouse gas emission
   D. large area deforestation

5. **Describe** how human activities can impact climate.

6. **Identify** the advantages and disadvantages of global climate models.

7. **Describe** two ways deforestation contributes to the greenhouse effect.

Interpret Graphics

8. **Determine Cause and Effect** Draw a graphic organizer like the one below to identify two ways burning fossil fuels impacts climate.

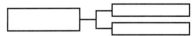

Critical Thinking

9. **Suggest** ways you can reduce greenhouse gas emissions.

10. **Assess** the effects of global warming in the area where you live.

Math Skills

Review — Math Practice

11. A 32-inch LCD flat-panel TV uses about 125 watts of electricity. If the screen size is increased to 40 inches, the TV uses 200 watts of electricity. What is the percent reduction of electricity if you use a 32-inch TV instead of a 40-inch TV?

Inquiry Lab

80 minutes

The greenhouse effect is a gas!

Materials

plastic wrap

2 jars with lids

sand

thermometer

desk lamp

stopwatch

rubber band

Safety

Human survival on Earth depends on the greenhouse effect. How can you model the greenhouse effect to help understand how it keeps Earth's temperature in balance?

Ask a Question

How will the temperature in a greenhouse compare to that of an open system when exposed to solar energy?

Make Observations

1. Read and complete a lab safety form.
2. Decide which type of container you think will make a good model of a greenhouse. Make two identical models.
3. Place equal amounts of sand in the bottom of each greenhouse.
4. Place a thermometer in each greenhouse in a position where you can read the temperature. Secure it on the wall of the container so you are not measuring the temperature of the sand.
5. Leave one container open, and close the other container.
6. Place the greenhouses under a light source—the Sun or a lamp. Have the light source the same distance from each greenhouse and at the same angle.
7. Read the starting temperature and then every 5–10 minutes for at least three readings. Record the temperatures in your Science Journal and organize them in a table like the one shown on the next page.

Form a Hypothesis

8. Think about some adjustments you could make to your greenhouses to model other components of the greenhouse effect. For example, translucent tops, or white tops, could represent materials that would reflect more light and thermal energy.
9. Based on your observations, form a hypothesis about what materials would most accurately model the greenhouse effect.

Chapter 14
EXTEND

Temperature (°C)			
	Reading 1	Reading 2	Reading 3
Greenhouse 1			
Greenhouse 2			

Test Your Hypothesis

10. Set up both greenhouse models in the same way for the hypothesis you are testing. Determine how many trials are sufficient for a valid conclusion. Graph your data to give a visual for your comparison.

Analyze and Conclude

11. Did thermal energy escape from either model? How does this compare to solar energy that reaches Earth and radiates back into the atmosphere?

12. If the greenhouse gases trap thermal energy and keep Earth's temperature warm enough, what would happen if they were not in the atmosphere?

13. If too much of a greenhouse gas, such as CO_2, entered the atmosphere, would the temperature rise?

14. **The Big Idea** If you could add water vapor or CO_2 to your model greenhouses to create an imbalance of greenhouse gases, would this affect the temperature of either system? Apply this to Earth's greenhouse gases.

Communicate Your Results

Discuss your findings with your group and organize your data. Share your graphs, models, and conclusions with the class. Explain why you chose certain materials and how these related directly to your hypothesis.

Now that you understand the importance of the function of the greenhouse effect, do further investigating into what happens when the balance of greenhouse gases changes. This could result in global warming, which can have a very negative impact on Earth and the atmosphere. Design an experiment that could show how global warming occurs.

Lab Tips

☑ Focus on one concept in designing your lab so you do not get confused with the complexities of materials and data.

☑ Do not add clouds to your greenhouse as part of your model. Clouds are condensed water; water vapor is a gas.

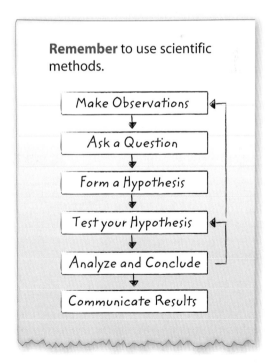

Chapter 14 Study Guide

Climate is the long-term average weather conditions that occur in an area. Living things have adaptations to the climate in which they live.

Key Concepts Summary

Lesson 1: Climates of Earth

- **Climate** is the long-term average weather conditions that occur in a particular region.
- Climate is affected by factors such as latitude, altitude, **rain shadows** on the downwind slope of mountains, vegetation, and the **specific heat** of water.
- Climate is classified based on precipitation, temperature, and native vegetation.

Vocabulary

climate p. 487
rain shadow p. 489
specific heat p. 489
microclimate p. 491

Lesson 2: Climate Cycles

- Over the past 4.6 billion years, climate on Earth has varied between **ice ages** and warm periods. **Interglacials** mark warm periods on Earth during or between ice ages.
- Earth's axis is tilted. This causes seasons as Earth revolves around the Sun.
- The **El Niño/Southern Oscillation** and **monsoons** are two climate patterns that result from interactions between oceans and the atmosphere.

ice age p. 496
interglacial p. 496
El Niño/Southern Oscillation p. 500
monsoon p. 501
drought p. 501

Lesson 3: Recent Climate Change

- Releasing carbon dioxide and aerosols into the atmosphere through burning fossil fuels and **deforestation** are two ways humans can affect climate change.
- Predictions about future climate change are made using computers and **global climate models.**

global warming p. 506
greenhouse gas p. 506
deforestation p. 507
global climate model p. 509

Study Guide

- Personal Tutor
- Vocabulary eGames
- Vocabulary eFlashcards

FOLDABLES Chapter Project

Assemble your lesson Foldables as shown to make a Chapter Project. Use the project to review what you have learned in this chapter.

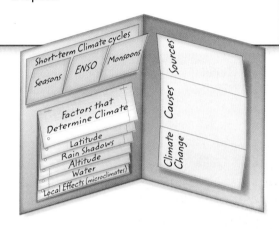

Use Vocabulary

1. A(n) _____ is an area of low rainfall on the downwind slope of a mountain.

2. Forests often have their own _____, with cooler temperatures than the surrounding countryside.

3. The lower _____ of land causes it to warm up faster than water.

4. A wind circulation pattern that changes direction with the seasons is a(n) _____.

5. Upwelling, trade winds, and air pressure patterns across the Pacific Ocean change during a(n) _____.

6. Earth's current _____ is called the Holocene Epoch.

7. A(n) _____ such as carbon dioxide absorbs Earth's infrared radiation and warms the atmosphere.

8. Additional CO_2 is added to the atmosphere when _____ of large land areas occurs.

Link Vocabulary and Key Concepts

Concepts in Motion Interactive Concept Map

Copy this concept map, and then use vocabulary terms from the previous page and other terms in this chapter to complete the concept map.

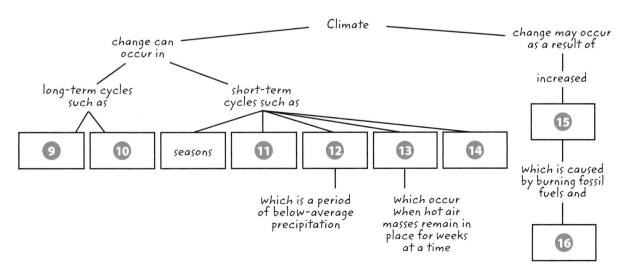

Chapter 14 Study Guide • **515**

Chapter 14 Review

Understand Key Concepts

1. The specific heat of water is _____ than the specific heat of land.
   A. higher
   B. lower
   C. less efficient
   D. more efficient

2. The graph below shows average monthly temperature and precipitation of an area over the course of a year.

   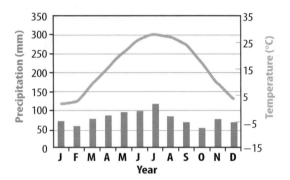

   Which is the most likely location of the area?
   A. in the middle of a large continent
   B. in the middle of the ocean
   C. near the North Pole
   D. on the coast of a large continent

3. Which are warm periods during or between ice ages?
   A. ENSO
   B. interglacials
   C. monsoons
   D. Pacific oscillations

4. Long-term climate cycles are caused by all of the following EXCEPT
   A. changes in ocean circulation.
   B. Earth's revolution of the Sun.
   C. the slow movement of the continents.
   D. variations in the shape of Earth's orbit.

5. A rain shadow is created by which factor that affects climate?
   A. a large body of water
   B. buildings and concrete
   C. latitude
   D. mountains

6. During which event do trade winds weaken and the usual pattern of pressure across the Pacific Ocean reverses?
   A. drought
   B. El Niño/Southern Oscillation event
   C. North Atlantic Oscillation event
   D. volcanic eruption

7. The picture below shows Earth as it revolves around the Sun.

   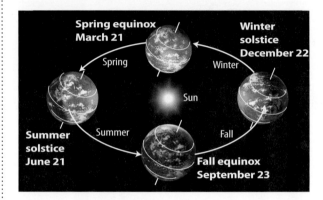

   Which season is it in the southern hemisphere in July?
   A. fall
   B. spring
   C. summer
   D. winter

8. Which is not a greenhouse gas?
   A. carbon dioxide
   B. methane
   C. oxygen
   D. water vapor

9. Which cools the climate by preventing sunlight from reaching Earth's surface?
   A. aerosols
   B. greenhouse gases
   C. lakes
   D. water vapor molecules

10. Which action can reduce greenhouse gas emissions?
    A. building houses on permafrost
    B. burning fossil fuels
    C. cutting down forests
    D. driving a hybrid vehicle

Chapter Review

Assessment — Online Test Practice

Critical Thinking

11 Hypothesize how the climate of your town would change if North America and Asia moved together and became one enormous continent.

12 Interpret Graphics Identify the factor that affects climate, as shown in this graph. How does this factor affect climate?

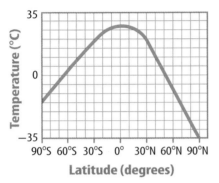

13 Diagram Draw a diagram that explains the changes that occur during an El Niño/Southern Oscillation event.

14 Evaluate which would cause more problems for your city or town: a drought, a heat wave, or a cold wave. Explain.

15 Recommend a life change you could make if the climate in your city were to change.

16 Formulate your opinion about the cause of global warming. Use facts to support your opinion.

17 Predict the effects of population increase on the climate where you live.

18 Compare how moisture affects the climates on either side of a mountain range.

Writing in Science

19 Write a short paragraph that describes a microclimate near your school or your home. What is the cause of the microclimate?

REVIEW THE BIG IDEA

20 What is climate? Explain what factors affect climate and give three examples of different types of climate.

21 Explain how life on Earth is affected by climate.

Math Skills — Math Practice

Use Percentages

22 Fred switches from a sport-utility vehicle that uses 800 gal of gasoline a year to a compact car that uses 450 gal.

  a. By what percent did Fred reduce the amount of gasoline used?

  b. If each gallon of gasoline released 20 pounds of CO_2, by what percent did Fred reduce the released CO_2?

23 Of the 186 billion tons of CO_2 that enter Earth's atmosphere each year from all sources, 6 billion tons are from human activity. If humans reduced their CO_2 production by half, what percentage decrease would it make in the total CO_2 entering the atmosphere?

Standardized Test Practice

Record your answers on the answer sheet provided by your teacher or on a sheet of paper.

Multiple Choice

1. Which is a drawback of a global climate model?
   A Its accuracy is nearly impossible to evaluate.
   B Its calculations are limited to specific regions.
   C Its predictions are short-term only.
   D Its results are difficult to interpret.

Use the diagram below to answer question 2.

2. What kind of climate would you expect to find at position 4?
   A mild
   B continental
   C tropical
   D dry

3. The difference in air temperature between a city and the surrounding rural area is an example of a(n)
   A inversion.
   B microclimate.
   C seasonal variation.
   D weather system.

4. Which does NOT help explain climate differences?
   A altitude
   B latitude
   C oceans
   D organisms

5. What is the primary cause of seasonal changes on Earth?
   A Earth's distance from the Sun
   B Earth's ocean currents
   C Earth's prevailing winds
   D Earth's tilt on its axis

Use the diagram below to answer question 6.

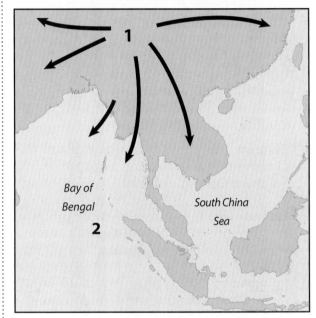

6. In the above diagram of the Asian winter monsoon, what does 1 represent?
   A high pressure
   B increased precipitation
   C low temperatures
   D wind speed

7. Climate is the _____ average weather conditions that occur in a particular region. Which completes the definition of *climate*?
   A global
   B long-term
   C mid-latitude
   D seasonal

Standardized Test Practice

Use the diagram below to answer question 8.

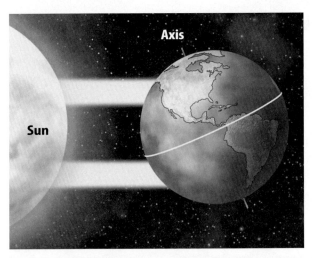

8 In the diagram above, what season is North America experiencing?

  A fall
  B spring
  C summer
  D winter

9 Which climate typically has warm summers, cold winters, and moderate precipitation?

  A continental
  B dry
  C polar
  D tropical

10 Which characterizes interglacials?

  A earthquakes
  B monsoons
  C precipitation
  D warmth

Constructed Response

Use the diagram below to answer question 11.

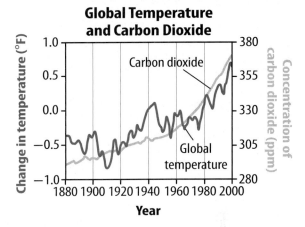

11 Compare the lines in the graph above. What does this graph suggest about the relationship between global temperature and atmospheric carbon dioxide?

Use the table below to answer questions 12 and 13.

Human Sources	Natural Sources

12 List two human and three natural sources of carbon dioxide. How do the listed human activities increase carbon dioxide levels in the atmosphere?

13 Which human activity listed in the table above also produces aerosols? What are two ways aerosols cool Earth?

NEED EXTRA HELP?												
If You Missed Question...	1	2	3	4	5	6	7	8	9	10	11	12	13
Go to Lesson...	3	1	1	1	2	2	1	2	1	2	3	3	3

Student Resources

For Students and Parents/Guardians

These resources are designed to help you achieve success in science. You will find useful information on laboratory safety, math skills, and science skills. In addition, science reference materials are found in the Reference Handbook. You'll find the information you need to learn and sharpen your skills in these resources.

Table of Contents

Science Skill Handbook SR-2
Scientific Methods SR-2
- Identify a Question SR-2
- Gather and Organize Information SR-2
- Form a Hypothesis SR-5
- Test the Hypothesis SR-6
- Collect Data SR-6
- Analyze the Data SR-9
- Draw Conclusions SR-10
- Communicate SR-10

Safety Symbols SR-11
Safety in the Science Laboratory SR-12
- General Safety Rules SR-12
- Prevent Accidents SR-12
- Laboratory Work SR-13
- Emergencies SR-13

Math Skill Handbook SR-14
Math Review SR-14
- Use Fractions SR-14
- Use Ratios SR-17
- Use Decimals SR-17
- Use Proportions SR-18
- Use Percentages SR-19
- Solve One-Step Equations SR-19
- Use Statistics SR-20
- Use Geometry SR-21

Science Application SR-24
- Measure in SI SR-24
- Dimensional Analysis SR-24
- Precision and Significant Digits SR-26
- Scientific Notation SR-26
- Make and Use Graphs SR-27

Foldables Handbook SR-29

Reference Handbook SR-40
- Periodic Table of the Elements SR-40
- Topographic Map Symbols SR-42
- Rocks SR-43
- Minerals SR-44
- Weather Map Symbols SR-46

Glossary G-2

Index I-2

Credits C-2

Science Skill Handbook

Scientific Methods

Scientists use an orderly approach called the scientific method to solve problems. This includes organizing and recording data so others can understand them. Scientists use many variations in this method when they solve problems.

Identify a Question

The first step in a scientific investigation or experiment is to identify a question to be answered or a problem to be solved. For example, you might ask which gasoline is the most efficient.

Gather and Organize Information

After you have identified your question, begin gathering and organizing information. There are many ways to gather information, such as researching in a library, interviewing those knowledgeable about the subject, and testing and working in the laboratory and field. Fieldwork is investigations and observations done outside of a laboratory.

Researching Information Before moving in a new direction, it is important to gather the information that already is known about the subject. Start by asking yourself questions to determine exactly what you need to know. Then you will look for the information in various reference sources, like the student is doing in **Figure 1**. Some sources may include textbooks, encyclopedias, government documents, professional journals, science magazines, and the Internet. Always list the sources of your information.

Figure 1 The Internet can be a valuable research tool.

Evaluate Sources of Information Not all sources of information are reliable. You should evaluate all of your sources of information, and use only those you know to be dependable. For example, if you are researching ways to make homes more energy efficient, a site written by the U.S. Department of Energy would be more reliable than a site written by a company that is trying to sell a new type of weatherproofing material. Also, remember that research always is changing. Consult the most current resources available to you. For example, a 1985 resource about saving energy would not reflect the most recent findings.

Sometimes scientists use data that they did not collect themselves, or conclusions drawn by other researchers. This data must be evaluated carefully. Ask questions about how the data were obtained, if the investigation was carried out properly, and if it has been duplicated exactly with the same results. Would you reach the same conclusion from the data? Only when you have confidence in the data can you believe it is true and feel comfortable using it.

SR-2 • Science Skill Handbook

Interpret Scientific Illustrations As you research a topic in science, you will see drawings, diagrams, and photographs to help you understand what you read. Some illustrations are included to help you understand an idea that you can't see easily by yourself, like the tiny particles in an atom in **Figure 2**. A drawing helps many people to remember details more easily and provides examples that clarify difficult concepts or give additional information about the topic you are studying. Most illustrations have labels or a caption to identify or to provide more information.

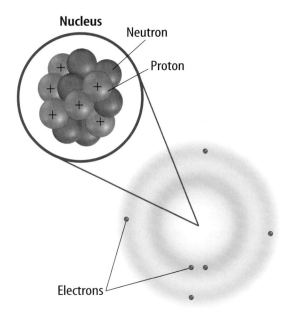

Figure 2 This drawing shows an atom of carbon with its six protons, six neutrons, and six electrons.

Concept Maps One way to organize data is to draw a diagram that shows relationships among ideas (or concepts). A concept map can help make the meanings of ideas and terms more clear, and help you understand and remember what you are studying. Concept maps are useful for breaking large concepts down into smaller parts, making learning easier.

Network Tree A type of concept map that not only shows a relationship, but how the concepts are related is a network tree, shown in **Figure 3**. In a network tree, the words are written in the ovals, while the description of the type of relationship is written across the connecting lines.

When constructing a network tree, write down the topic and all major topics on separate pieces of paper or notecards. Then arrange them in order from general to specific. Branch the related concepts from the major concept and describe the relationship on the connecting line. Continue to more specific concepts until finished.

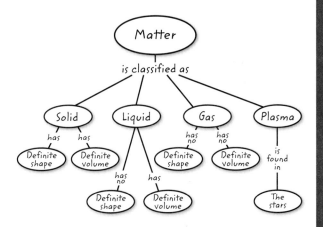

Figure 3 A network tree shows how concepts or objects are related.

Events Chain Another type of concept map is an events chain. Sometimes called a flow chart, it models the order or sequence of items. An events chain can be used to describe a sequence of events, the steps in a procedure, or the stages of a process.

When making an events chain, first find the one event that starts the chain. This event is called the initiating event. Then, find the next event and continue until the outcome is reached, as shown in **Figure 4** on the next page.

Science Skill Handbook • **SR-3**

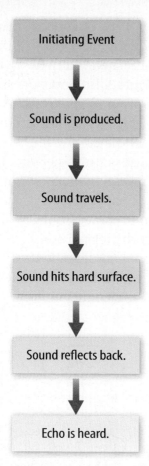

Figure 4 Events-chain concept maps show the order of steps in a process or event. This concept map shows how a sound makes an echo.

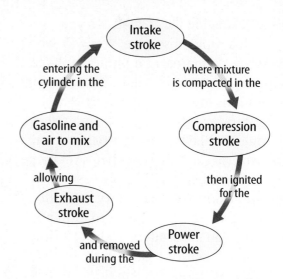

Figure 5 A cycle map shows events that occur in a cycle.

Cycle Map A specific type of events chain is a cycle map. It is used when the series of events do not produce a final outcome, but instead relate back to the beginning event, such as in **Figure 5**. Therefore, the cycle repeats itself.

To make a cycle map, first decide what event is the beginning event. This is also called the initiating event. Then list the next events in the order that they occur, with the last event relating back to the initiating event. Words can be written between the events that describe what happens from one event to the next. The number of events in a cycle map can vary, but usually contain three or more events.

Spider Map A type of concept map that you can use for brainstorming is the spider map. When you have a central idea, you might find that you have a jumble of ideas that relate to it but are not necessarily clearly related to each other. The spider map on sound in **Figure 6** shows that if you write these ideas outside the main concept, then you can begin to separate and group unrelated terms so they become more useful.

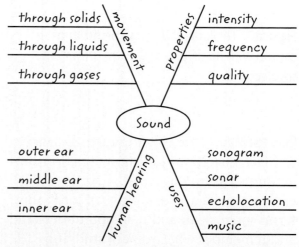

Figure 6 A spider map allows you to list ideas that relate to a central topic but not necessarily to one another.

SR-4 • Science Skill Handbook

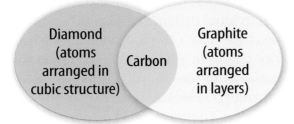

Figure 7 This Venn diagram compares and contrasts two substances made from carbon.

Venn Diagram To illustrate how two subjects compare and contrast you can use a Venn diagram. You can see the characteristics that the subjects have in common and those that they do not, shown in **Figure 7**.

To create a Venn diagram, draw two overlapping ovals that are big enough to write in. List the characteristics unique to one subject in one oval, and the characteristics of the other subject in the other oval. The characteristics in common are listed in the overlapping section.

Make and Use Tables One way to organize information so it is easier to understand is to use a table. Tables can contain numbers, words, or both.

To make a table, list the items to be compared in the first column and the characteristics to be compared in the first row. The title should clearly indicate the content of the table, and the column or row heads should be clear. Notice that in **Table 1** the units are included.

Table 1 Recyclables Collected During Week			
Day of Week	Paper (kg)	Aluminum (kg)	Glass (kg)
Monday	5.0	4.0	12.0
Wednesday	4.0	1.0	10.0
Friday	2.5	2.0	10.0

Make a Model One way to help you better understand the parts of a structure, the way a process works, or to show things too large or small for viewing is to make a model. For example, an atomic model made of a plastic-ball nucleus and chenille stem electron shells can help you visualize how the parts of an atom relate to each other. Other types of models can be devised on a computer or represented by equations.

Form a Hypothesis

A possible explanation based on previous knowledge and observations is called a hypothesis. After researching gasoline types and recalling previous experiences in your family's car you form a hypothesis—our car runs more efficiently because we use premium gasoline. To be valid, a hypothesis has to be something you can test by using an investigation.

Predict When you apply a hypothesis to a specific situation, you predict something about that situation. A prediction makes a statement in advance, based on prior observation, experience, or scientific reasoning. People use predictions to make everyday decisions. Scientists test predictions by performing investigations. Based on previous observations and experiences, you might form a prediction that cars are more efficient with premium gasoline. The prediction can be tested in an investigation.

Design an Experiment A scientist needs to make many decisions before beginning an investigation. Some of these include: how to carry out the investigation, what steps to follow, how to record the data, and how the investigation will answer the question. It also is important to address any safety concerns.

Test the Hypothesis

Now that you have formed your hypothesis, you need to test it. Using an investigation, you will make observations and collect data, or information. This data might either support or not support your hypothesis. Scientists collect and organize data as numbers and descriptions.

Follow a Procedure In order to know what materials to use, as well as how and in what order to use them, you must follow a procedure. **Figure 8** shows a procedure you might follow to test your hypothesis.

Procedure

Step 1	Use regular gasoline for two weeks.
Step 2	Record the number of kilometers between fill-ups and the amount of gasoline used.
Step 3	Switch to premium gasoline for two weeks.
Step 4	Record the number of kilometers between fill-ups and the amount of gasoline used.

Figure 8 A procedure tells you what to do step-by-step.

Identify and Manipulate Variables and Controls In any experiment, it is important to keep everything the same except for the item you are testing. The one factor you change is called the independent variable. The change that results is the dependent variable. Make sure you have only one independent variable, to assure yourself of the cause of the changes you observe in the dependent variable. For example, in your gasoline experiment the type of fuel is the independent variable. The dependent variable is the efficiency.

Many experiments also have a control—an individual instance or experimental subject for which the independent variable is not changed. You can then compare the test results to the control results. To design a control you can have two cars of the same type. The control car uses regular gasoline for four weeks. After you are done with the test, you can compare the experimental results to the control results.

Collect Data

Whether you are carrying out an investigation or a short observational experiment, you will collect data, as shown in **Figure 9**. Scientists collect data as numbers and descriptions and organize them in specific ways.

Observe Scientists observe items and events, then record what they see. When they use only words to describe an observation, it is called qualitative data. Scientists' observations also can describe how much there is of something. These observations use numbers, as well as words, in the description and are called quantitative data. For example, if a sample of the element gold is described as being "shiny and very dense" the data are qualitative. Quantitative data on this sample of gold might include "a mass of 30 g and a density of 19.3 g/cm^3."

Figure 9 Collecting data is one way to gather information directly.

Figure 10 Record data neatly and clearly so it is easy to understand.

When you make observations you should examine the entire object or situation first, and then look carefully for details. It is important to record observations accurately and completely. Always record your notes immediately as you make them, so you do not miss details or make a mistake when recording results from memory. Never put unidentified observations on scraps of paper. Instead they should be recorded in a notebook, like the one in **Figure 10.** Write your data neatly so you can easily read it later. At each point in the experiment, record your observations and label them. That way, you will not have to determine what the figures mean when you look at your notes later. Set up any tables that you will need to use ahead of time, so you can record any observations right away. Remember to avoid bias when collecting data by not including personal thoughts when you record observations. Record only what you observe.

Estimate Scientific work also involves estimating. To estimate is to make a judgment about the size or the number of something without measuring or counting. This is important when the number or size of an object or population is too large or too difficult to accurately count or measure.

Sample Scientists may use a sample or a portion of the total number as a type of estimation. To sample is to take a small, representative portion of the objects or organisms of a population for research. By making careful observations or manipulating variables within that portion of the group, information is discovered and conclusions are drawn that might apply to the whole population. A poorly chosen sample can be unrepresentative of the whole. If you were trying to determine the rainfall in an area, it would not be best to take a rainfall sample from under a tree.

Measure You use measurements every day. Scientists also take measurements when collecting data. When taking measurements, it is important to know how to use measuring tools properly. Accuracy also is important.

Length To measure length, the distance between two points, scientists use meters. Smaller measurements might be measured in centimeters or millimeters.

Length is measured using a metric ruler or meterstick. When using a metric ruler, line up the 0-cm mark with the end of the object being measured and read the number of the unit where the object ends. Look at the metric ruler shown in **Figure 11.** The centimeter lines are the long, numbered lines, and the shorter lines are millimeter lines. In this instance, the length would be 4.50 cm.

Figure 11 This metric ruler has centimeter and millimeter divisions.

Science Skill Handbook • **SR-7**

Mass The SI unit for mass is the kilogram (kg). Scientists can measure mass using units formed by adding metric prefixes to the unit gram (g), such as milligram (mg). To measure mass, you might use a triple-beam balance similar to the one shown in **Figure 12.** The balance has a pan on one side and a set of beams on the other side. Each beam has a rider that slides on the beam.

When using a triple-beam balance, place an object on the pan. Slide the largest rider along its beam until the pointer drops below zero. Then move it back one notch. Repeat the process for each rider proceeding from the larger to smaller until the pointer swings an equal distance above and below the zero point. Sum the masses on each beam to find the mass of the object. Move all riders back to zero when finished.

Instead of putting materials directly on the balance, scientists often take a tare of a container. A tare is the mass of a container into which objects or substances are placed for measuring their masses. To find the mass of objects or substances, find the mass of a clean container. Remove the container from the pan, and place the object or substances in the container. Find the mass of the container with the materials in it. Subtract the mass of the empty container from the mass of the filled container to find the mass of the materials you are using.

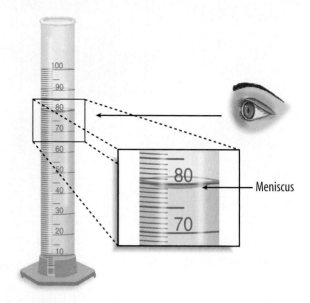

Figure 13 Graduated cylinders measure liquid volume.

Liquid Volume To measure liquids, the unit used is the liter. When a smaller unit is needed, scientists might use a milliliter. Because a milliliter takes up the volume of a cube measuring 1 cm on each side it also can be called a cubic centimeter ($cm^3 = cm \times cm \times cm$).

You can use beakers and graduated cylinders to measure liquid volume. A graduated cylinder, shown in **Figure 13,** is marked from bottom to top in milliliters. In lab, you might use a 10-mL graduated cylinder or a 100-mL graduated cylinder. When measuring liquids, notice that the liquid has a curved surface. Look at the surface at eye level, and measure the bottom of the curve. This is called the meniscus. The graduated cylinder in **Figure 13** contains 79.0 mL, or 79.0 cm^3, of a liquid.

Temperature Scientists often measure temperature using the Celsius scale. Pure water has a freezing point of 0°C and boiling point of 100°C. The unit of measurement is degrees Celsius. Two other scales often used are the Fahrenheit and Kelvin scales.

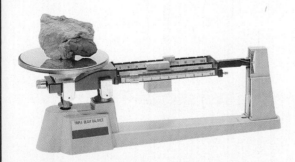

Figure 12 A triple-beam balance is used to determine the mass of an object.

SR-8 • Science Skill Handbook

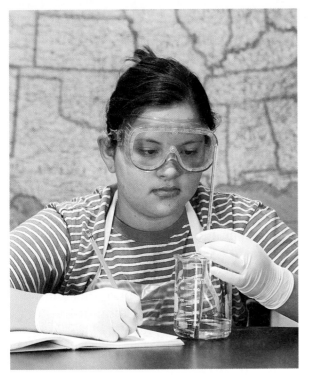

Figure 14 A thermometer measures the temperature of an object.

Scientists use a thermometer to measure temperature. Most thermometers in a laboratory are glass tubes with a bulb at the bottom end containing a liquid such as colored alcohol. The liquid rises or falls with a change in temperature. To read a glass thermometer like the thermometer in **Figure 14,** rotate it slowly until a red line appears. Read the temperature where the red line ends.

Form Operational Definitions An operational definition defines an object by how it functions, works, or behaves. For example, when you are playing hide and seek and a tree is home base, you have created an operational definition for a tree.

Objects can have more than one operational definition. For example, a ruler can be defined as a tool that measures the length of an object (how it is used). It can also be a tool with a series of marks used as a standard when measuring (how it works).

Analyze the Data

To determine the meaning of your observations and investigation results, you will need to look for patterns in the data. Then you must think critically to determine what the data mean. Scientists use several approaches when they analyze the data they have collected and recorded. Each approach is useful for identifying specific patterns.

Interpret Data The word *interpret* means "to explain the meaning of something." When analyzing data from an experiment, try to find out what the data show. Identify the control group and the test group to see whether changes in the independent variable have had an effect. Look for differences in the dependent variable between the control and test groups.

Classify Sorting objects or events into groups based on common features is called classifying. When classifying, first observe the objects or events to be classified. Then select one feature that is shared by some members in the group, but not by all. Place those members that share that feature in a subgroup. You can classify members into smaller and smaller subgroups based on characteristics. Remember that when you classify, you are grouping objects or events for a purpose. Keep your purpose in mind as you select the features to form groups and subgroups.

Compare and Contrast Observations can be analyzed by noting the similarities and differences between two or more objects or events that you observe. When you look at objects or events to see how they are similar, you are comparing them. Contrasting is looking for differences in objects or events.

Science Skill Handbook • **SR-9**

Recognize Cause and Effect A cause is a reason for an action or condition. The effect is that action or condition. When two events happen together, it is not necessarily true that one event caused the other. Scientists must design a controlled investigation to recognize the exact cause and effect.

Draw Conclusions

When scientists have analyzed the data they collected, they proceed to draw conclusions about the data. These conclusions are sometimes stated in words similar to the hypothesis that you formed earlier. They may confirm a hypothesis, or lead you to a new hypothesis.

Infer Scientists often make inferences based on their observations. An inference is an attempt to explain observations or to indicate a cause. An inference is not a fact, but a logical conclusion that needs further investigation. For example, you may infer that a fire has caused smoke. Until you investigate, however, you do not know for sure.

Apply When you draw a conclusion, you must apply those conclusions to determine whether the data supports the hypothesis. If your data do not support your hypothesis, it does not mean that the hypothesis is wrong. It means only that the result of the investigation did not support the hypothesis. Maybe the experiment needs to be redesigned, or some of the initial observations on which the hypothesis was based were incomplete or biased. Perhaps more observation or research is needed to refine your hypothesis. A successful investigation does not always come out the way you originally predicted.

Avoid Bias Sometimes a scientific investigation involves making judgments. When you make a judgment, you form an opinion. It is important to be honest and not to allow any expectations of results to bias your judgments. This is important throughout the entire investigation, from researching to collecting data to drawing conclusions.

Communicate

The communication of ideas is an important part of the work of scientists. A discovery that is not reported will not advance the scientific community's understanding or knowledge. Communication among scientists also is important as a way of improving their investigations.

Scientists communicate in many ways, from writing articles in journals and magazines that explain their investigations and experiments, to announcing important discoveries on television and radio. Scientists also share ideas with colleagues on the Internet or present them as lectures, like the student is doing in **Figure 15**.

Figure 15 A student communicates to his peers about his investigation.

These safety symbols are used in laboratory and field investigations in this book to indicate possible hazards. Learn the meaning of each symbol and refer to this page often. *Remember to wash your hands thoroughly after completing lab procedures.*

PROTECTIVE EQUIPMENT
Do not begin any lab without the proper protection equipment.

 GOGGLES Proper eye protection must be worn when performing or observing science activities that involve items or conditions as listed below.

 APRON Wear an approved apron when using substances that could stain, wet, or destroy cloth.

 SOAP Wash hands with soap and water before removing goggles and after all lab activities.

 GLOVES Wear gloves when working with biological materials, chemicals, animals, or materials that can stain or irritate hands.

LABORATORY HAZARDS

Symbols	Potential Hazards	Precaution	Response
DISPOSAL	contamination of classroom or environment due to improper disposal of materials such as chemicals and live specimens	• DO NOT dispose of hazardous materials in the sink or trash can. • Dispose of wastes as directed by your teacher.	• If hazardous materials are disposed of improperly, notify your teacher immediately.
EXTREME TEMPERATURE	skin burns due to extremely hot or cold materials such as hot glass, liquids, or metals; liquid nitrogen; dry ice	• Use proper protective equipment, such as hot mitts and/or tongs, when handling objects with extreme temperatures.	• If injury occurs, notify your teacher immediately.
SHARP OBJECTS	punctures or cuts from sharp objects such as razor blades, pins, scalpels, and broken glass	• Handle glassware carefully to avoid breakage. • Walk with sharp objects pointed downward, away from you and others.	• If broken glass or injury occurs, notify your teacher immediately.
ELECTRICAL	electric shock or skin burn due to improper grounding, short circuits, liquid spills, or exposed wires	• Check condition of wires and apparatus for fraying or uninsulated wires, and broken or cracked equipment. • Use only GFCI-protected outlets	• DO NOT attempt to fix electrical problems. Notify your teacher immediately.
CHEMICAL	skin irritation or burns, breathing difficulty, and/or poisoning due to touching, swallowing, or inhalation of chemicals such as acids, bases, bleach, metal compounds, iodine, poinsettias, pollen, ammonia, acetone, nail polish remover, heated chemicals, mothballs, and any other chemicals labeled or known to be dangerous	• Wear proper protective equipment such as goggles, apron, and gloves when using chemicals. • Ensure proper room ventilation or use a fume hood when using materials that produce fumes. • NEVER smell fumes directly. • NEVER taste or eat any material in the laboratory.	• If contact occurs, immediately flush affected area with water and notify your teacher. • If a spill occurs, leave the area immediately and notify your teacher.
FLAMMABLE	unexpected fire due to liquids or gases that ignite easily such as rubbing alcohol	• Avoid open flames, sparks, or heat when flammable liquids are present.	• If a fire occurs, leave the area immediately and notify your teacher.
OPEN FLAME	burns or fire due to open flame from matches, Bunsen burners, or burning materials	• Tie back loose hair and clothing. • Keep flame away from all materials. • Follow teacher instructions when lighting and extinguishing flames. • Use proper protection, such as hot mitts or tongs, when handling hot objects.	• If a fire occurs, leave the area immediately and notify your teacher.
ANIMAL SAFETY	injury to or from laboratory animals	• Wear proper protective equipment such as gloves, apron, and goggles when working with animals. • Wash hands after handling animals.	• If injury occurs, notify your teacher immediately.
BIOLOGICAL	infection or adverse reaction due to contact with organisms such as bacteria, fungi, and biological materials such as blood, animal or plant materials	• Wear proper protective equipment such as gloves, goggles, and apron when working with biological materials. • Avoid skin contact with an organism or any part of the organism. • Wash hands after handling organisms.	• If contact occurs, wash the affected area and notify your teacher immediately.
FUME	breathing difficulties from inhalation of fumes from substances such as ammonia, acetone, nail polish remover, heated chemicals, and mothballs	• Wear goggles, apron, and gloves. • Ensure proper room ventilation or use a fume hood when using substances that produce fumes. • NEVER smell fumes directly.	• If a spill occurs, leave area and notify your teacher immediately.
IRRITANT	irritation of skin, mucous membranes, or respiratory tract due to materials such as acids, bases, bleach, pollen, mothballs, steel wool, and potassium permanganate	• Wear goggles, apron, and gloves. • Wear a dust mask to protect against fine particles.	• If skin contact occurs, immediately flush the affected area with water and notify your teacher.
RADIOACTIVE	excessive exposure from alpha, beta, and gamma particles	• Remove gloves and wash hands with soap and water before removing remainder of protective equipment.	• If cracks or holes are found in the container, notify your teacher immediately.

Safety in the Science Laboratory

Introduction to Science Safety

The science laboratory is a safe place to work if you follow standard safety procedures. Being responsible for your own safety helps to make the entire laboratory a safer place for everyone. When performing any lab, read and apply the caution statements and safety symbol listed at the beginning of the lab.

General Safety Rules

1. Complete the *Lab Safety Form* or other safety contract BEFORE starting any science lab.
2. Study the procedure. Ask your teacher any questions. Be sure you understand safety symbols shown on the page.
3. Notify your teacher about allergies or other health conditions that can affect your participation in a lab.
4. Learn and follow use and safety procedures for your equipment. If unsure, ask your teacher.

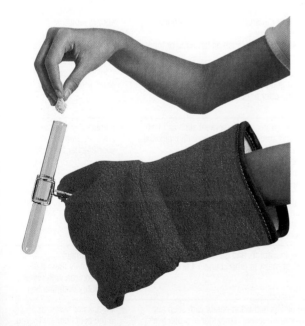

5. Never eat, drink, chew gum, apply cosmetics, or do any personal grooming in the lab. Never use lab glassware as food or drink containers. Keep your hands away from your face and mouth.
6. Know the location and proper use of the safety shower, eye wash, fire blanket, and fire alarm.

Prevent Accidents

1. Use the safety equipment provided to you. Goggles and a safety apron should be worn during investigations.
2. Do NOT use hair spray, mousse, or other flammable hair products. Tie back long hair and tie down loose clothing.
3. Do NOT wear sandals or other open-toed shoes in the lab.
4. Remove jewelry on hands and wrists. Loose jewelry, such as chains and long necklaces, should be removed to prevent them from getting caught in equipment.
5. Do not taste any substances or draw any material into a tube with your mouth.
6. Proper behavior is expected in the lab. Practical jokes and fooling around can lead to accidents and injury.
7. Keep your work area uncluttered.

Laboratory Work

1. Collect and carry all equipment and materials to your work area before beginning a lab.
2. Remain in your own work area unless given permission by your teacher to leave it.

3. Always slant test tubes away from yourself and others when heating them, adding substances to them, or rinsing them.
4. If instructed to smell a substance in a container, hold the container a short distance away and fan vapors toward your nose.
5. Do NOT substitute other chemicals/substances for those in the materials list unless instructed to do so by your teacher.
6. Do NOT take any materials or chemicals outside of the laboratory.
7. Stay out of storage areas unless instructed to be there and supervised by your teacher.

Laboratory Cleanup

1. Turn off all burners, water, and gas, and disconnect all electrical devices.
2. Clean all pieces of equipment and return all materials to their proper places.
3. Dispose of chemicals and other materials as directed by your teacher. Place broken glass and solid substances in the proper containers. Never discard materials in the sink.
4. Clean your work area.
5. Wash your hands with soap and water thoroughly BEFORE removing your goggles.

Emergencies

1. Report any fire, electrical shock, glassware breakage, spill, or injury, no matter how small, to your teacher immediately. Follow his or her instructions.
2. If your clothing should catch fire, STOP, DROP, and ROLL. If possible, smother it with the fire blanket or get under a safety shower. NEVER RUN.
3. If a fire should occur, turn off all gas and leave the room according to established procedures.
4. In most instances, your teacher will clean up spills. Do NOT attempt to clean up spills unless you are given permission and instructions to do so.
5. If chemicals come into contact with your eyes or skin, notify your teacher immediately. Use the eyewash, or flush your skin or eyes with large quantities of water.
6. The fire extinguisher and first-aid kit should only be used by your teacher unless it is an extreme emergency and you have been given permission.
7. If someone is injured or becomes ill, only a professional medical provider or someone certified in first aid should perform first-aid procedures.

Math Skill Handbook

Math Review

Use Fractions

A fraction compares a part to a whole. In the fraction $\frac{2}{3}$, the 2 represents the part and is the numerator. The 3 represents the whole and is the denominator.

Reduce Fractions To reduce a fraction, you must find the largest factor that is common to both the numerator and the denominator, the greatest common factor (GCF). Divide both numbers by the GCF. The fraction has then been reduced, or it is in its simplest form.

Example

Twelve of the 20 chemicals in the science lab are in powder form. What fraction of the chemicals used in the lab are in powder form?

Step 1 Write the fraction.

$$\frac{part}{whole} = \frac{12}{20}$$

Step 2 To find the GCF of the numerator and denominator, list all of the factors of each number.

Factors of 12: 1, 2, 3, 4, 6, 12 (the numbers that divide evenly into 12)

Factors of 20: 1, 2, 4, 5, 10, 20 (the numbers that divide evenly into 20)

Step 3 List the common factors.

1, 2, 4

Step 4 Choose the greatest factor in the list. The GCF of 12 and 20 is 4.

Step 5 Divide the numerator and denominator by the GCF.

$$\frac{12 \div 4}{20 \div 4} = \frac{3}{5}$$

In the lab, $\frac{3}{5}$ of the chemicals are in powder form.

Practice Problem At an amusement park, 66 of 90 rides have a height restriction. What fraction of the rides, in its simplest form, has a height restriction?

Add and Subtract Fractions with Like Denominators To add or subtract fractions with the same denominator, add or subtract the numerators and write the sum or difference over the denominator. After finding the sum or difference, find the simplest form for your fraction.

Example 1

In the forest outside your house, $\frac{1}{8}$ of the animals are rabbits, $\frac{3}{8}$ are squirrels, and the remainder are birds and insects. How many are mammals?

Step 1 Add the numerators.

$$\frac{1}{8} + \frac{3}{8} = \frac{(1+3)}{8} = \frac{4}{8}$$

Step 2 Find the GCF.

$$\frac{4}{8} \text{ (GCF, 4)}$$

Step 3 Divide the numerator and denominator by the GCF.

$$\frac{4 \div 4}{8 \div 4} = \frac{1}{2}$$

$\frac{1}{2}$ of the animals are mammals.

Example 2

If $\frac{7}{16}$ of the Earth is covered by freshwater, and $\frac{1}{16}$ of that is in glaciers, how much freshwater is not frozen?

Step 1 Subtract the numerators.

$$\frac{7}{16} - \frac{1}{16} = \frac{(7-1)}{16} = \frac{6}{16}$$

Step 2 Find the GCF.

$$\frac{6}{16} \text{ (GCF, 2)}$$

Step 3 Divide the numerator and denominator by the GCF.

$$\frac{6 \div 2}{16 \div 2} = \frac{3}{8}$$

$\frac{3}{8}$ of the freshwater is not frozen.

Practice Problem A bicycle rider is riding at a rate of 15 km/h for $\frac{4}{9}$ of his ride, 10 km/h for $\frac{2}{9}$ of his ride, and 8 km/h for the remainder of the ride. How much of his ride is he riding at a rate greater than 8 km/h?

Add and Subtract Fractions with Unlike Denominators To add or subtract fractions with unlike denominators, first find the least common denominator (LCD). This is the smallest number that is a common multiple of both denominators. Rename each fraction with the LCD, and then add or subtract. Find the simplest form if necessary.

Example 1

A chemist makes a paste that is $\frac{1}{2}$ table salt (NaCl), $\frac{1}{3}$ sugar ($C_6H_{12}O_6$), and the remainder is water (H_2O). How much of the paste is a solid?

Step 1 Find the LCD of the fractions.

$\frac{1}{2} + \frac{1}{3}$ (LCD, 6)

Step 2 Rename each numerator and each denominator with the LCD.

Step 3 Add the numerators.

$\frac{3}{6} + \frac{2}{6} = \frac{(3+2)}{6} = \frac{5}{6}$

$\frac{5}{6}$ of the paste is a solid.

Example 2

The average precipitation in Grand Junction, CO, is $\frac{7}{10}$ inch in November, and $\frac{3}{5}$ inch in December. What is the total average precipitation?

Step 1 Find the LCD of the fractions.

$\frac{7}{10} + \frac{3}{5}$ (LCD, 10)

Step 2 Rename each numerator and each denominator with the LCD.

Step 3 Add the numerators.

$\frac{7}{10} + \frac{6}{10} = \frac{(7+6)}{10} = \frac{13}{10}$

$\frac{13}{10}$ inches total precipitation, or $1\frac{3}{10}$ inches.

Practice Problem On an electric bill, about $\frac{1}{8}$ of the energy is from solar energy and about $\frac{1}{10}$ is from wind power. How much of the total bill is from solar energy and wind power combined?

Example 3

In your body, $\frac{7}{10}$ of your muscle contractions are involuntary (cardiac and smooth muscle tissue). Smooth muscle makes $\frac{3}{15}$ of your muscle contractions. How many of your muscle contractions are made by cardiac muscle?

Step 1 Find the LCD of the fractions.

$\frac{7}{10} - \frac{3}{15}$ (LCD, 30)

Step 2 Rename each numerator and each denominator with the LCD.

$\frac{7 \times 3}{10 \times 3} = \frac{21}{30}$

$\frac{3 \times 2}{15 \times 2} = \frac{6}{30}$

Step 3 Subtract the numerators.

$\frac{21}{30} - \frac{6}{30} = \frac{(21-6)}{30} = \frac{15}{30}$

Step 4 Find the GCF.

$\frac{15}{30}$ (GCF, 15)

$\frac{1}{2}$

$\frac{1}{2}$ of all muscle contractions are cardiac muscle.

Example 4

Tony wants to make cookies that call for $\frac{3}{4}$ of a cup of flour, but he only has $\frac{1}{3}$ of a cup. How much more flour does he need?

Step 1 Find the LCD of the fractions.

$\frac{3}{4} - \frac{1}{3}$ (LCD, 12)

Step 2 Rename each numerator and each denominator with the LCD.

$\frac{3 \times 3}{4 \times 3} = \frac{9}{12}$

$\frac{1 \times 4}{3 \times 4} = \frac{4}{12}$

Step 3 Subtract the numerators.

$\frac{9}{12} - \frac{4}{12} = \frac{(9-4)}{12} = \frac{5}{12}$

$\frac{5}{12}$ of a cup of flour

Practice Problem Using the information provided to you in Example 3 above, determine how many muscle contractions are voluntary (skeletal muscle).

Multiply Fractions To multiply with fractions, multiply the numerators and multiply the denominators. Find the simplest form if necessary.

> **Example**
>
> Multiply $\frac{3}{5}$ by $\frac{1}{3}$.
>
> **Step 1** Multiply the numerators and denominators.
>
> $$\frac{3}{5} \times \frac{1}{3} = \frac{(3 \times 1)}{(5 \times 3)} \quad \frac{3}{15}$$
>
> **Step 2** Find the GCF.
>
> $\frac{3}{15}$ (GCF, 3)
>
> **Step 3** Divide the numerator and denominator by the GCF.
>
> $$\frac{3 \div 3}{15 \div 3} = \frac{1}{5}$$
>
> $\frac{3}{5}$ multiplied by $\frac{1}{3}$ is $\frac{1}{5}$.

Practice Problem Multiply $\frac{3}{14}$ by $\frac{5}{16}$.

Find a Reciprocal Two numbers whose product is 1 are called multiplicative inverses, or reciprocals.

> **Example**
>
> Find the reciprocal of $\frac{3}{8}$.
>
> **Step 1** Inverse the fraction by putting the denominator on top and the numerator on the bottom.
>
> $\frac{8}{3}$
>
> The reciprocal of $\frac{3}{8}$ is $\frac{8}{3}$.

Practice Problem Find the reciprocal of $\frac{4}{9}$.

Divide Fractions To divide one fraction by another fraction, multiply the dividend by the reciprocal of the divisor. Find the simplest form if necessary.

> **Example 1**
>
> Divide $\frac{1}{9}$ by $\frac{1}{3}$.
>
> **Step 1** Find the reciprocal of the divisor.
>
> The reciprocal of $\frac{1}{3}$ is $\frac{3}{1}$.
>
> **Step 2** Multiply the dividend by the reciprocal of the divisor.
>
> $$\frac{\frac{1}{9}}{\frac{1}{3}} = \frac{1}{9} \times \frac{3}{1} = \frac{(1 \times 3)}{(9 \times 1)} = \frac{3}{9}$$
>
> **Step 3** Find the GCF.
>
> $\frac{3}{9}$ (GCF, 3)
>
> **Step 4** Divide the numerator and denominator by the GCF.
>
> $$\frac{3 \div 3}{9 \div 3} = \frac{1}{3}$$
>
> $\frac{1}{9}$ divided by $\frac{1}{3}$ is $\frac{1}{3}$.

> **Example 2**
>
> Divide $\frac{3}{5}$ by $\frac{1}{4}$.
>
> **Step 1** Find the reciprocal of the divisor.
>
> The reciprocal of $\frac{1}{4}$ is $\frac{4}{1}$.
>
> **Step 2** Multiply the dividend by the reciprocal of the divisor.
>
> $$\frac{\frac{3}{5}}{\frac{1}{4}} = \frac{3}{5} \times \frac{4}{1} = \frac{(3 \times 4)}{(5 \times 1)} = \frac{12}{5}$$
>
> $\frac{3}{5}$ divided by $\frac{1}{4}$ is $\frac{12}{5}$ or $2\frac{2}{5}$.

Practice Problem Divide $\frac{3}{11}$ by $\frac{7}{10}$.

Use Ratios

When you compare two numbers by division, you are using a ratio. Ratios can be written 3 to 5, 3:5, or $\frac{3}{5}$. Ratios, like fractions, also can be written in simplest form.

Ratios can represent one type of probability, called odds. This is a ratio that compares the number of ways a certain outcome occurs to the number of possible outcomes. For example, if you flip a coin 100 times, what are the odds that it will come up heads? There are two possible outcomes, heads or tails, so the odds of coming up heads are 50:100. Another way to say this is that 50 out of 100 times the coin will come up heads. In its simplest form, the ratio is 1:2.

Example 1

A chemical solution contains 40 g of salt and 64 g of baking soda. What is the ratio of salt to baking soda as a fraction in simplest form?

Step 1 Write the ratio as a fraction.

$$\frac{salt}{baking\ soda} = \frac{40}{64}$$

Step 2 Express the fraction in simplest form. The GCF of 40 and 64 is 8.

$$\frac{40}{64} = \frac{40 \div 8}{64 \div 8} = \frac{5}{8}$$

The ratio of salt to baking soda in the sample is 5:8.

Example 2

Sean rolls a 6-sided die 6 times. What are the odds that the side with a 3 will show?

Step 1 Write the ratio as a fraction.

$$\frac{number\ of\ sides\ with\ a\ 3}{number\ of\ possible\ sides} = \frac{1}{6}$$

Step 2 Multiply by the number of attempts.

$$\frac{1}{6} \times 6\ attempts = \frac{6}{6}\ attempts = 1\ attempt$$

1 attempt out of 6 will show a 3.

Practice Problem Two metal rods measure 100 cm and 144 cm in length. What is the ratio of their lengths in simplest form?

Use Decimals

A fraction with a denominator that is a power of ten can be written as a decimal. For example, 0.27 means $\frac{27}{100}$. The decimal point separates the ones place from the tenths place.

Any fraction can be written as a decimal using division. For example, the fraction $\frac{5}{8}$ can be written as a decimal by dividing 5 by 8. Written as a decimal, it is 0.625.

Add or Subtract Decimals When adding and subtracting decimals, line up the decimal points before carrying out the operation.

Example 1

Find the sum of 47.68 and 7.80.

Step 1 Line up the decimal places when you write the numbers.

```
  47.68
+  7.80
```

Step 2 Add the decimals.

```
  47.68
+  7.80
  55.48
```

The sum of 47.68 and 7.80 is 55.48.

Example 2

Find the difference of 42.17 and 15.85.

Step 1 Line up the decimal places when you write the number.

```
  42.17
- 15.85
```

Step 2 Subtract the decimals.

```
  42.17
- 15.85
  26.32
```

The difference of 42.17 and 15.85 is 26.32.

Practice Problem Find the sum of 1.245 and 3.842.

Multiply Decimals To multiply decimals, multiply the numbers like numbers without decimal points. Count the decimal places in each factor. The product will have the same number of decimal places as the sum of the decimal places in the factors.

> **Example**
>
> Multiply 2.4 by 5.9.
>
> **Step 1** Multiply the factors like two whole numbers.
>
> $24 \times 59 = 1416$
>
> **Step 2** Find the sum of the number of decimal places in the factors. Each factor has one decimal place, for a sum of two decimal places.
>
> **Step 3** The product will have two decimal places.
>
> 14.16
>
> The product of 2.4 and 5.9 is 14.16.

Practice Problem Multiply 4.6 by 2.2.

Divide Decimals When dividing decimals, change the divisor to a whole number. To do this, multiply both the divisor and the dividend by the same power of ten. Then place the decimal point in the quotient directly above the decimal point in the dividend. Then divide as you do with whole numbers.

> **Example**
>
> Divide 8.84 by 3.4.
>
> **Step 1** Multiply both factors by 10.
>
> $3.4 \times 10 = 34, 8.84 \times 10 = 88.4$
>
> **Step 2** Divide 88.4 by 34.
>
> $$\begin{array}{r} 2.6 \\ 34\overline{)88.4} \\ -68 \\ \hline 204 \\ -204 \\ \hline 0 \end{array}$$
>
> 8.84 divided by 3.4 is 2.6.

Practice Problem Divide 75.6 by 3.6.

Use Proportions

An equation that shows that two ratios are equivalent is a proportion. The ratios $\frac{2}{4}$ and $\frac{5}{10}$ are equivalent, so they can be written as $\frac{2}{4} = \frac{5}{10}$. This equation is a proportion.

When two ratios form a proportion, the cross products are equal. To find the cross products in the proportion $\frac{2}{4} = \frac{5}{10}$, multiply the 2 and the 10, and the 4 and the 5. Therefore $2 \times 10 = 4 \times 5$, or $20 = 20$.

Because you know that both ratios are equal, you can use cross products to find a missing term in a proportion. This is known as solving the proportion.

> **Example**
>
> The heights of a tree and a pole are proportional to the lengths of their shadows. The tree casts a shadow of 24 m when a 6-m pole casts a shadow of 4 m. What is the height of the tree?
>
> **Step 1** Write a proportion.
>
> $$\frac{\text{height of tree}}{\text{height of pole}} = \frac{\text{length of tree's shadow}}{\text{length of pole's shadow}}$$
>
> **Step 2** Substitute the known values into the proportion. Let h represent the unknown value, the height of the tree.
>
> $\frac{h}{6} \times \frac{24}{4}$
>
> **Step 3** Find the cross products.
>
> $h \times 4 = 6 \times 24$
>
> **Step 4** Simplify the equation.
>
> $4h \times 144$
>
> **Step 5** Divide each side by 4.
>
> $\frac{4h}{4} \times \frac{144}{4}$
>
> $h = 36$
>
> The height of the tree is 36 m.

Practice Problem The ratios of the weights of two objects on the Moon and on Earth are in proportion. A rock weighing 3 N on the Moon weighs 18 N on Earth. How much would a rock that weighs 5 N on the Moon weigh on Earth?

Use Percentages

The word *percent* means "out of one hundred." It is a ratio that compares a number to 100. Suppose you read that 77 percent of Earth's surface is covered by water. That is the same as reading that the fraction of Earth's surface covered by water is $\frac{77}{100}$. To express a fraction as a percent, first find the equivalent decimal for the fraction. Then, multiply the decimal by 100 and add the percent symbol.

Example 1

Express $\frac{13}{20}$ as a percent.

Step 1 Find the equivalent decimal for the fraction.

$$\begin{array}{r} 0.65 \\ 20\overline{)13.00} \\ \underline{12\ 0} \\ 1\ 00 \\ \underline{1\ 00} \\ 0 \end{array}$$

Step 2 Rewrite the fraction $\frac{13}{20}$ as 0.65.

Step 3 Multiply 0.65 by 100 and add the % symbol.

$$0.65 \times 100 = 65 = 65\%$$

So, $\frac{13}{20} = 65\%$.

This also can be solved as a proportion.

Example 2

Express $\frac{13}{20}$ as a percent.

Step 1 Write a proportion.

$$\frac{13}{20} = \frac{x}{100}$$

Step 2 Find the cross products.

$$1300 = 20x$$

Step 3 Divide each side by 20.

$$\frac{1300}{20} = \frac{20x}{20}$$
$$65\% = x$$

Practice Problem In one year, 73 of 365 days were rainy in one city. What percent of the days in that city were rainy?

Solve One-Step Equations

A statement that two expressions are equal is an equation. For example, $A = B$ is an equation that states that A is equal to B.

An equation is solved when a variable is replaced with a value that makes both sides of the equation equal. To make both sides equal the inverse operation is used. Addition and subtraction are inverses, and multiplication and division are inverses.

Example 1

Solve the equation $x - 10 = 35$.

Step 1 Find the solution by adding 10 to each side of the equation.

$$x - 10 = 35$$
$$x - 10 + 10 = 35 - 10$$
$$x = 45$$

Step 2 Check the solution.

$$x - 10 = 35$$
$$45 - 10 = 35$$
$$35 = 35$$

Both sides of the equation are equal, so $x = 45$.

Example 2

In the formula $a = bc$, find the value of c if $a = 20$ and $b = 2$.

Step 1 Rearrange the formula so the unknown value is by itself on one side of the equation by dividing both sides by b.

$$a = bc$$
$$\frac{a}{b} = \frac{bc}{b}$$
$$\frac{a}{b} = c$$

Step 2 Replace the variables a and b with the values that are given.

$$\frac{a}{b} = c$$
$$\frac{20}{2} = c$$
$$10 = c$$

Step 3 Check the solution.

$$a = bc$$
$$20 = 2 \times 10$$
$$20 = 20$$

Both sides of the equation are equal, so $c = 10$ is the solution when $a = 20$ and $b = 2$.

Practice Problem In the formula $h = gd$, find the value of d if $g = 12.3$ and $h = 17.4$.

Use Statistics

The branch of mathematics that deals with collecting, analyzing, and presenting data is statistics. In statistics, there are three common ways to summarize data with a single number—the mean, the median, and the mode.

The **mean** of a set of data is the arithmetic average. It is found by adding the numbers in the data set and dividing by the number of items in the set.

The **median** is the middle number in a set of data when the data are arranged in numerical order. If there were an even number of data points, the median would be the mean of the two middle numbers.

The **mode** of a set of data is the number or item that appears most often.

Another number that often is used to describe a set of data is the range. The **range** is the difference between the largest number and the smallest number in a set of data.

Example

The speeds (in m/s) for a race car during five different time trials are 39, 37, 44, 36, and 44.

To find the mean:

Step 1 Find the sum of the numbers.

39 + 37 + 44 + 36 + 44 = 200

Step 2 Divide the sum by the number of items, which is 5.

200 ÷ 5 = 40

The mean is 40 m/s.

To find the median:

Step 1 Arrange the measures from least to greatest.

36, 37, 39, 44, 44

Step 2 Determine the middle measure.

36, 37, <u>39</u>, 44, 44

The median is 39 m/s.

To find the mode:

Step 1 Group the numbers that are the same together.

44, 44, 36, 37, 39

Step 2 Determine the number that occurs most in the set.

<u>44, 44,</u> 36, 37, 39

The mode is 44 m/s.

To find the range:

Step 1 Arrange the measures from greatest to least.

44, 44, 39, 37, 36

Step 2 Determine the greatest and least measures in the set.

<u>44,</u> 44, 39, 37, <u>36</u>

Step 3 Find the difference between the greatest and least measures.

44 − 36 = 8

The range is 8 m/s.

Practice Problem Find the mean, median, mode, and range for the data set 8, 4, 12, 8, 11, 14, 16.

A **frequency table** shows how many times each piece of data occurs, usually in a survey. **Table 1** below shows the results of a student survey on favorite color.

Table 1 Student Color Choice		
Color	Tally	Frequency
red	IIII	4
blue	|||||	5
black	II	2
green	III	3
purple	||||| II	7
yellow	||||| I	6

Based on the frequency table data, which color is the favorite?

Use Geometry

The branch of mathematics that deals with the measurement, properties, and relationships of points, lines, angles, surfaces, and solids is called geometry.

Perimeter The **perimeter** (P) is the distance around a geometric figure. To find the perimeter of a rectangle, add the length and width and multiply that sum by two, or $2(l + w)$. To find perimeters of irregular figures, add the length of the sides.

Example 1

Find the perimeter of a rectangle that is 3 m long and 5 m wide.

Step 1 You know that the perimeter is 2 times the sum of the width and length.

$P = 2(3 \text{ m} + 5 \text{ m})$

Step 2 Find the sum of the width and length.

$P = 2(8 \text{ m})$

Step 3 Multiply by 2.

$P = 16 \text{ m}$

The perimeter is 16 m.

Example 2

Find the perimeter of a shape with sides measuring 2 cm, 5 cm, 6 cm, 3 cm.

Step 1 You know that the perimeter is the sum of all the sides.

$P = 2 + 5 + 6 + 3$

Step 2 Find the sum of the sides.

$P = 2 + 5 + 6 + 3$

$P = 16$

The perimeter is 16 cm.

Practice Problem Find the perimeter of a rectangle with a length of 18 m and a width of 7 m.

Practice Problem Find the perimeter of a triangle measuring 1.6 cm by 2.4 cm by 2.4 cm.

Area of a Rectangle The **area** (A) is the number of square units needed to cover a surface. To find the area of a rectangle, multiply the length times the width, or $l \times w$. When finding area, the units also are multiplied. Area is given in square units.

Example

Find the area of a rectangle with a length of 1 cm and a width of 10 cm.

Step 1 You know that the area is the length multiplied by the width.

$A = (1 \text{ cm} \times 10 \text{ cm})$

Step 2 Multiply the length by the width. Also multiply the units.

$A = 10 \text{ cm}^2$

The area is 10 cm^2.

Practice Problem Find the area of a square whose sides measure 4 m.

Area of a Triangle To find the area of a triangle, use the formula:

$A = \frac{1}{2}(\text{base} \times \text{height})$

The base of a triangle can be any of its sides. The height is the perpendicular distance from a base to the opposite endpoint, or vertex.

Example

Find the area of a triangle with a base of 18 m and a height of 7 m.

Step 1 You know that the area is $\frac{1}{2}$ the base times the height.

$A = \frac{1}{2}(18 \text{ m} \times 7 \text{ m})$

Step 2 Multiply $\frac{1}{2}$ by the product of 18×7. Multiply the units.

$A = \frac{1}{2}(126 \text{ m}^2)$

$A = 63 \text{ m}^2$

The area is 63 m^2.

Practice Problem Find the area of a triangle with a base of 27 cm and a height of 17 cm.

Circumference of a Circle The **diameter** (*d*) of a circle is the distance across the circle through its center, and the **radius** (r) is the distance from the center to any point on the circle. The radius is half of the diameter. The distance around the circle is called the **circumference** (C). The formula for finding the circumference is:

$C = 2\pi r$ or $C = \pi d$

The circumference divided by the diameter is always equal to 3.1415926… This nonterminating and nonrepeating number is represented by the Greek letter π (pi). An approximation often used for π is 3.14.

Example 1

Find the circumference of a circle with a radius of 3 m.

Step 1 You know the formula for the circumference is 2 times the radius times π.

$C = 2\pi(3)$

Step 2 Multiply 2 times the radius.

$C = 6\pi$

Step 3 Multiply by π.

$C \approx 19$ m

The circumference is about 19 m.

Example 2

Find the circumference of a circle with a diameter of 24.0 cm.

Step 1 You know the formula for the circumference is the diameter times π.

$C = \pi(24.0)$

Step 2 Multiply the diameter by π.

$C \approx 75.4$ cm

The circumference is about 75.4 cm.

Practice Problem Find the circumference of a circle with a radius of 19 cm.

Area of a Circle The formula for the area of a circle is: $A = \pi r^2$

Example 1

Find the area of a circle with a radius of 4.0 cm.

Step 1 $A = \pi(4.0)^2$

Step 2 Find the square of the radius.

$A = 16\pi$

Step 3 Multiply the square of the radius by π.

$A \approx 50$ cm^2

The area of the circle is about 50 cm^2.

Example 2

Find the area of a circle with a radius of 225 m.

Step 1 $A = \pi(225)^2$

Step 2 Find the square of the radius.

$A = 50625\pi$

Step 3 Multiply the square of the radius by π.

$A \approx 159043.1$

The area of the circle is about 159043.1 m^2.

Example 3

Find the area of a circle whose diameter is 20.0 mm.

Step 1 Remember that the radius is half of the diameter.

$A = \pi\left(\dfrac{20.0}{2}\right)^2$

Step 2 Find the radius.

$A = \pi(10.0)^2$

Step 3 Find the square of the radius.

$A = 100\pi$

Step 4 Multiply the square of the radius by π.

$A \approx 314$ mm^2

The area of the circle is about 314 mm^2.

Practice Problem Find the area of a circle with a radius of 16 m.

Volume The measure of space occupied by a solid is the **volume** (V). To find the volume of a rectangular solid multiply the length times width times height, or $V = l \times w \times h$. It is measured in cubic units, such as cubic centimeters (cm^3).

Example

Find the volume of a rectangular solid with a length of 2.0 m, a width of 4.0 m, and a height of 3.0 m.

Step 1 You know the formula for volume is the length times the width times the height.

$V = 2.0 \text{ m} \times 4.0 \text{ m} \times 3.0 \text{ m}$

Step 2 Multiply the length times the width times the height.

$V = 24 \text{ m}^3$

The volume is 24 m^3.

Practice Problem Find the volume of a rectangular solid that is 8 m long, 4 m wide, and 4 m high.

To find the volume of other solids, multiply the area of the base times the height.

Example 1

Find the volume of a solid that has a triangular base with a length of 8.0 m and a height of 7.0 m. The height of the entire solid is 15.0 m.

Step 1 You know that the base is a triangle, and the area of a triangle is $\frac{1}{2}$ the base times the height, and the volume is the area of the base times the height.

$V = \left[\frac{1}{2}(b \times h)\right] \times 15$

Step 2 Find the area of the base.

$V = \left[\frac{1}{2}(8 \times 7)\right] \times 15$

$V = \left(\frac{1}{2} \times 56\right) \times 15$

Step 3 Multiply the area of the base by the height of the solid.

$V = 28 \times 15$

$V = 420 \text{ m}^3$

The volume is 420 m^3.

Example 2

Find the volume of a cylinder that has a base with a radius of 12.0 cm, and a height of 21.0 cm.

Step 1 You know that the base is a circle, and the area of a circle is the square of the radius times π, and the volume is the area of the base times the height.

$V = (\pi r^2) \times 21$

$V = (\pi 12^2) \times 21$

Step 2 Find the area of the base.

$V = 144\pi \times 21$

$V = 452 \times 21$

Step 3 Multiply the area of the base by the height of the solid.

$V \approx 9{,}500 \text{ cm}^3$

The volume is about 9,500 cm^3.

Example 3

Find the volume of a cylinder that has a diameter of 15 mm and a height of 4.8 mm.

Step 1 You know that the base is a circle with an area equal to the square of the radius times π. The radius is one-half the diameter. The volume is the area of the base times the height.

$V = (\pi r^2) \times 4.8$

$V = \left[\pi\left(\frac{1}{2} \times 15\right)^2\right] \times 4.8$

$V = (\pi 7.5^2) \times 4.8$

Step 2 Find the area of the base.

$V = 56.25\pi \times 4.8$

$V \approx 176.71 \times 4.8$

Step 3 Multiply the area of the base by the height of the solid.

$V \approx 848.2$

The volume is about 848.2 mm^3.

Practice Problem Find the volume of a cylinder with a diameter of 7 cm in the base and a height of 16 cm.

Math Skill Handbook • **SR-23**

Science Applications

Measure in SI

The metric system of measurement was developed in 1795. A modern form of the metric system, called the International System (SI), was adopted in 1960 and provides the standard measurements that all scientists around the world can understand.

The SI system is convenient because unit sizes vary by powers of 10. Prefixes are used to name units. Look at **Table 2** for some common SI prefixes and their meanings.

Table 2 Common SI Prefixes

Prefix	Symbol	Meaning	
kilo–	k	1,000	thousandth
hecto–	h	100	hundred
deka–	da	10	ten
deci–	d	0.1	tenth
centi–	c	0.01	hundreth
milli–	m	0.001	thousandth

Example

How many grams equal one kilogram?

Step 1 Find the prefix *kilo–* in **Table 2.**

Step 2 Using **Table 2,** determine the meaning of *kilo–*. According to the table, it means 1,000. When the prefix *kilo–* is added to a unit, it means that there are 1,000 of the units in a "kilounit."

Step 3 Apply the prefix to the units in the question. The units in the question are grams. There are 1,000 grams in a kilogram.

Practice Problem Is a milligram larger or smaller than a gram? How many of the smaller units equal one larger unit? What fraction of the larger unit does one smaller unit represent?

Dimensional Analysis

Convert SI Units In science, quantities such as length, mass, and time sometimes are measured using different units. A process called dimensional analysis can be used to change one unit of measure to another. This process involves multiplying your starting quantity and units by one or more conversion factors. A conversion factor is a ratio equal to one and can be made from any two equal quantities with different units. If 1,000 mL equal 1 L then two ratios can be made.

$$\frac{1{,}000 \text{ mL}}{1 \text{ L}} = \frac{1 \text{ L}}{1{,}000 \text{ mL}} = 1$$

One can convert between units in the SI system by using the equivalents in **Table 2** to make conversion factors.

Example

How many cm are in 4 m?

Step 1 Write conversion factors for the units given. From **Table 2,** you know that 100 cm = 1 m. The conversion factors are

$$\frac{100 \text{ cm}}{1 \text{ m}} \text{ and } \frac{1 \text{ m}}{100 \text{ cm}}$$

Step 2 Decide which conversion factor to use. Select the factor that has the units you are converting from (m) in the denominator and the units you are converting to (cm) in the numerator.

$$\frac{100 \text{ cm}}{1 \text{ m}}$$

Step 3 Multiply the starting quantity and units by the conversion factor. Cancel the starting units with the units in the denominator. There are 400 cm in 4 m.

$$4 \cancel{\text{m}} = \frac{100 \text{ cm}}{1 \cancel{\text{m}}} = 400 \text{ cm}$$

Practice Problem How many milligrams are in one kilogram? (Hint: You will need to use two conversion factors from **Table 2.**)

Table 3 Unit System Equivalents

Type of Measurement	Equivalent
Length	1 in = 2.54 cm 1 yd = 0.91 m 1 mi = 1.61 km
Mass and weight*	1 oz = 28.35 g 1 lb = 0.45 kg 1 ton (short) = 0.91 tonnes (metric tons) 1 lb = 4.45 N
Volume	1 in^3 = 16.39 cm^3 1 qt = 0.95 L 1 gal = 3.78 L
Area	1 in^2 = 6.45 cm^2 1 yd^2 = 0.83 m^2 1 mi^2 = 2.59 km^2 1 acre = 0.40 hectares
Temperature	$°C = \frac{(°F - 32)}{1.8}$ $K = °C + 273$

*Weight is measured in standard Earth gravity.

Convert Between Unit Systems Table 3 gives a list of equivalents that can be used to convert between English and SI units.

Example

If a meterstick has a length of 100 cm, how long is the meterstick in inches?

Step 1 Write the conversion factors for the units given. From **Table 3,** 1 in = 2.54 cm.

$$\frac{1 \text{ in}}{2.54 \text{ cm}} \text{ and } \frac{2.54 \text{ cm}}{1 \text{ in}}$$

Step 2 Determine which conversion factor to use. You are converting from cm to in. Use the conversion factor with cm on the bottom.

$$\frac{1 \text{ in}}{2.54 \text{ cm}}$$

Step 3 Multiply the starting quantity and units by the conversion factor. Cancel the starting units with the units in the denominator. Round your answer to the nearest tenth.

$$100 \text{ cm} \times \frac{1 \text{ in}}{2.54 \text{ cm}} = 39.37 \text{ in}$$

The meterstick is about 39.4 in long.

Practice Problem 1 A book has a mass of 5 lb. What is the mass of the book in kg?

Practice Problem 2 Use the equivalent for in and cm (1 in = 2.54 cm) to show how 1 in^3 ≈ 16.39 cm^3.

Math Skill Handbook • **SR-25**

Precision and Significant Digits

When you make a measurement, the value you record depends on the precision of the measuring instrument. This precision is represented by the number of significant digits recorded in the measurement. When counting the number of significant digits, all digits are counted except zeros at the end of a number with no decimal point such as 2,050, and zeros at the beginning of a decimal such as 0.03020. When adding or subtracting numbers with different precision, round the answer to the smallest number of decimal places of any number in the sum or difference. When multiplying or dividing, the answer is rounded to the smallest number of significant digits of any number being multiplied or divided.

Example

The lengths 5.28 and 5.2 are measured in meters. Find the sum of these lengths and record your answer using the correct number of significant digits.

Step 1 Find the sum.

    5.28 m   2 digits after the decimal
+ 5.2 m   1 digit after the decimal
10.48 m

Step 2 Round to one digit after the decimal because the least number of digits after the decimal of the numbers being added is 1.

The sum is 10.5 m.

Practice Problem 1 How many significant digits are in the measurement 7,071,301 m? How many significant digits are in the measurement 0.003010 g?

Practice Problem 2 Multiply 5.28 and 5.2 using the rule for multiplying and dividing. Record the answer using the correct number of significant digits.

Scientific Notation

Many times numbers used in science are very small or very large. Because these numbers are difficult to work with scientists use scientific notation. To write numbers in scientific notation, move the decimal point until only one non-zero digit remains on the left. Then count the number of places you moved the decimal point and use that number as a power of ten. For example, the average distance from the Sun to Mars is 227,800,000,000 m. In scientific notation, this distance is 2.278×10^{11} m. Because you moved the decimal point to the left, the number is a positive power of ten.

The mass of an electron is about 0.000 000 000 000 000 000 000 000 000 000 911 kg. Expressed in scientific notation, this mass is 9.11×10^{-31} kg. Because the decimal point was moved to the right, the number is a negative power of ten.

Example

Earth is 149,600,000 km from the Sun. Express this in scientific notation.

Step 1 Move the decimal point until one non-zero digit remains on the left.

1.496 000 00

Step 2 Count the number of decimal places you have moved. In this case, eight.

Step 2 Show that number as a power of ten, 10^8.

Earth is 1.496×10^8 km from the Sun.

Practice Problem 1 How many significant digits are in 149,600,000 km? How many significant digits are in 1.496×10^8 km?

Practice Problem 2 Parts used in a high performance car must be measured to 7×10^{-6} m. Express this number as a decimal.

Practice Problem 3 A CD is spinning at 539 revolutions per minute. Express this number in scientific notation.

Make and Use Graphs

Data in tables can be displayed in a graph—a visual representation of data. Common graph types include line graphs, bar graphs, and circle graphs.

Line Graph A line graph shows a relationship between two variables that change continuously. The independent variable is changed and is plotted on the x-axis. The dependent variable is observed, and is plotted on the y-axis.

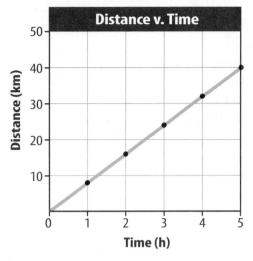

Figure 8 This line graph shows the relationship between distance and time during a bicycle ride.

Example

Draw a line graph of the data below from a cyclist in a long-distance race.

Table 4 Bicycle Race Data

Time (h)	Distance (km)
0	0
1	8
2	16
3	24
4	32
5	40

Step 1 Determine the x-axis and y-axis variables. Time varies independently of distance and is plotted on the x-axis. Distance is dependent on time and is plotted on the y-axis.

Step 2 Determine the scale of each axis. The x-axis data ranges from 0 to 5. The y-axis data ranges from 0 to 50.

Step 3 Using graph paper, draw and label the axes. Include units in the labels.

Step 4 Draw a point at the intersection of the time value on the x-axis and corresponding distance value on the y-axis. Connect the points and label the graph with a title, as shown in **Figure 8**.

Practice Problem A puppy's shoulder height is measured during the first year of her life. The following measurements were collected: (3 mo, 52 cm), (6 mo, 72 cm), (9 mo, 83 cm), (12 mo, 86 cm). Graph this data.

Find a Slope The slope of a straight line is the ratio of the vertical change, rise, to the horizontal change, run.

$$\text{Slope} = \frac{\text{vertical change (rise)}}{\text{horizontal change (run)}} = \frac{\text{change in } y}{\text{change in } x}$$

Example

Find the slope of the graph in **Figure 8**.

Step 1 You know that the slope is the change in y divided by the change in x.

$$\text{Slope} = \frac{\text{change in } y}{\text{change in } x}$$

Step 2 Determine the data points you will be using. For a straight line, choose the two sets of points that are the farthest apart.

$$\text{Slope} = \frac{(40 - 0) \text{ km}}{(5 - 0) \text{ h}}$$

Step 3 Find the change in y and x.

$$\text{Slope} = \frac{40 \text{ km}}{5 \text{ h}}$$

Step 4 Divide the change in y by the change in x.

$$\text{Slope} = \frac{8 \text{ km}}{\text{h}}$$

The slope of the graph is 8 km/h.

Math Skill Handbook • **SR-27**

Bar Graph To compare data that does not change continuously you might choose a bar graph. A bar graph uses bars to show the relationships between variables. The *x*-axis variable is divided into parts. The parts can be numbers such as years, or a category such as a type of animal. The *y*-axis is a number and increases continuously along the axis.

Example

A recycling center collects 4.0 kg of aluminum on Monday, 1.0 kg on Wednesday, and 2.0 kg on Friday. Create a bar graph of this data.

Step 1 Select the *x*-axis and *y*-axis variables. The measured numbers (the masses of aluminum) should be placed on the *y*-axis. The variable divided into parts (collection days) is placed on the *x*-axis.

Step 2 Create a graph grid like you would for a line graph. Include labels and units.

Step 3 For each measured number, draw a vertical bar above the *x*-axis value up to the *y*-axis value. For the first data point, draw a vertical bar above Monday up to 4.0 kg.

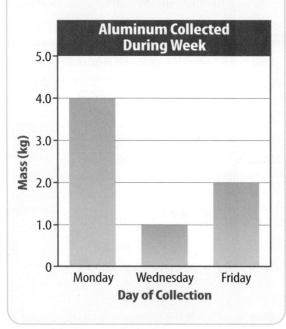

Practice Problem Draw a bar graph of the gases in air: 78% nitrogen, 21% oxygen, 1% other gases.

Circle Graph To display data as parts of a whole, you might use a circle graph. A circle graph is a circle divided into sections that represent the relative size of each piece of data. The entire circle represents 100%, half represents 50%, and so on.

Example

Air is made up of 78% nitrogen, 21% oxygen, and 1% other gases. Display the composition of air in a circle graph.

Step 1 Multiply each percent by 360° and divide by 100 to find the angle of each section in the circle.

$$78\% \times \frac{360°}{100} = 280.8°$$

$$21\% \times \frac{360°}{100} = 75.6°$$

$$1\% \times \frac{360°}{100} = 3.6°$$

Step 2 Use a compass to draw a circle and to mark the center of the circle. Draw a straight line from the center to the edge of the circle.

Step 3 Use a protractor and the angles you calculated to divide the circle into parts. Place the center of the protractor over the center of the circle and line the base of the protractor over the straight line.

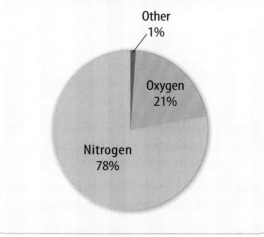

Practice Problem Draw a circle graph to represent the amount of aluminum collected during the week shown in the bar graph to the left.

Foldables® Handbook

Student Study Guides & Instructions
By Dinah Zike

1. You will find suggestions for Study Guides, also known as Foldables or books, in each chapter lesson and as a final project. Look at the end of the chapter to determine the project format and glue the Foldables in place as you progress through the chapter lessons.

2. Creating the Foldables or books is simple and easy to do by using copy paper, art paper, and internet printouts. Photocopies of maps, diagrams, or your own illustrations may also be used for some of the Foldables. Notebook paper is the most common source of material for study guides and 83% of all Foldables are created from it. When folded to make books, notebook paper Foldables easily fit into 11" × 17" or 12" × 18" chapter projects with space left over. Foldables made using photocopy paper are slightly larger and they fit into Projects, but snugly. Use the least amount of glue, tape, and staples needed to assemble the Foldables.

3. Seven of the Foldables can be made using either small or large paper. When 11" × 17" or 12" × 18" paper is used, these become projects for housing smaller Foldables. Project format boxes are located within the instructions to remind you of this option.

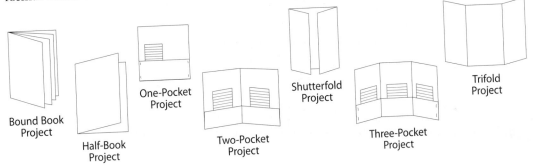

Bound Book Project Half-Book Project One-Pocket Project Two-Pocket Project Shutterfold Project Three-Pocket Project Trifold Project

4. Use one-gallon self-locking plastic bags to store your projects. Place strips of two-inch clear tape along the left, long side of the bag and punch holes through the taped edge. Cut the bottom corners off the bag so it will not hold air. Store this Project Portfolio inside a three-hole binder. To store a large collection of project bags, use a giant laundry-soap box. Holes can be punched in some of the Foldable Projects so they can be stored in a three-hole binder without using a plastic bag. Punch holes in the pocket books before gluing or stapling the pocket.

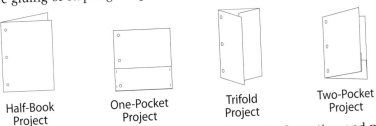

Half-Book Project One-Pocket Project Trifold Project Two-Pocket Project

5. Maximize the use of the projects by collecting additional information and placing it on the back of the project and other unused spaces of the large Foldables.

Half-Book Foldable® By Dinah Zike

Step 1 Fold a sheet of notebook or copy paper in half.

Label the exterior tab and use the inside space to write information.

PROJECT FORMAT
Use 11" × 17" or 12" × 18" paper on the horizontal axis to make a large project book.

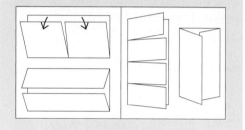

Variations
Paper can be folded horizontally, like a *hamburger* or vertically, like a *hot dog*.

C Half-books can be folded so that one side is ½ inch longer than the other side. A title or question can be written on the extended tab.

Worksheet Foldable or Folded Book® By Dinah Zike

Step 1 Make a half-book (see above) using work sheets, internet print-outs, diagrams, or maps.

Step 2 Fold it in half again.

Variations

A This folded sheet as a small book with two pages can be used for comparing and contrasting, cause and effect, or other skills.

B When the sheet of paper is open, the four sections can be used separately or used collectively to show sequences or steps.

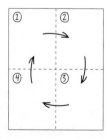

Two-Tab and Concept-Map Foldable® By Dinah Zike

Step 1 Fold a sheet of notebook or copy paper in half vertically or horizontally.

Step 2 Fold it in half again, as shown.

Step 3 Unfold once and cut along the fold line or valley of the top flap to make two flaps.

Variations

A Concept maps can be made by leaving a ½ inch tab at the top when folding the paper in half. Use arrows and labels to relate topics to the primary concept.

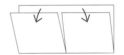

B Use two sheets of paper to make multiple page tab books. Glue or staple books together at the top fold.

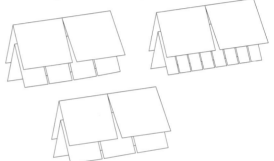

Three-Quarter Foldable® By Dinah Zike

Step 1 Make a two-tab book (see above) and cut the left tab off at the top of the fold line.

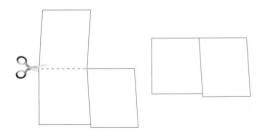

Variations

A Use this book to draw a diagram or a map on the exposed left tab. Write questions about the illustration on the top right tab and provide complete answers on the space under the tab.

B Compose a self-test using multiple choice answers for your questions. Include the correct answer with three wrong responses. The correct answers can be written on the back of the book or upside down on the bottom of the inside page.

Three-Tab Foldable® By Dinah Zike

Step 1 Fold a sheet of paper in half horizontally.

Step 2 Fold into thirds.

Step 3 Unfold and cut along the folds of the top flap to make three sections.

Variations

A Before cutting the three tabs draw a Venn diagram across the front of the book.

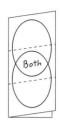

B Make a space to use for titles or concept maps by leaving a ½ inch tab at the top when folding the paper in half.

Four-Tab Foldable® By Dinah Zike

Step 1 Fold a sheet of paper in half horizontally.

Step 2 Fold in half and then fold each half as shown below.

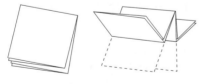

Step 3 Unfold and cut along the fold lines of the top flap to make four tabs.

Variations

A Make a space to use for titles or concept maps by leaving a ½ inch tab at the top when folding the paper in half.

B Use the book on the vertical axis, with or without an extended tab.

Folding Fifths for a Foldable® By Dinah Zike

Step 1 Fold a sheet of paper in half horizontally.

Step 2 Fold again so one-third of the paper is exposed and two-thirds are covered.

Step 3 Fold the two-thirds section in half.

Step 4 Fold the one-third section, a single thickness, backward to make a fold line.

Variations

A Unfold and cut along the fold lines to make five tabs.

B Make a five-tab book with a ½ inch tab at the top (see two-tab instructions).

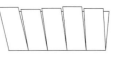

C Use 11" × 17" or 12" × 18" paper and fold into fifths for a five-column and/or row table or chart.

Folded Table or Chart, and Trifold Foldable® By Dinah Zike

Step 1 Fold a sheet of paper in the required number of vertical columns for the table or chart.

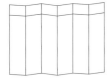

Step 2 Fold the horizontal rows needed for the table or chart.

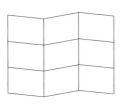

PROJECT FORMAT
Use 11" × 17" or 12" × 18" paper and fold it to make a large trifold project book or larger tables and charts.

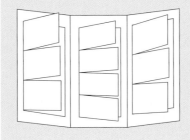

Variations

A Make a trifold by folding the paper into thirds vertically or horizontally.

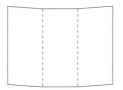

B Make a trifold book. Unfold it and draw a Venn diagram on the inside.

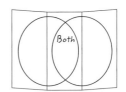

Two or Three-Pockets Foldable® By Dinah Zike

Step 1 Fold up the long side of a horizontal sheet of paper about 5 cm.

Step 2 Fold the paper in half.

Step 3 Open the paper and glue or staple the outer edges to make two compartments.

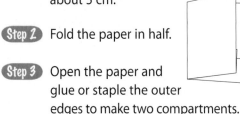

Variations

A Make a multi-page booklet by gluing several pocket books together.

B Make a three-pocket book by using a trifold (see previous instructions).

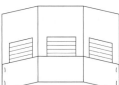

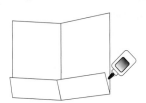

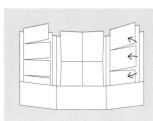

PROJECT FORMAT
Use 11" × 17" or 12" × 18" paper and fold it horizontally to make a large multi-pocket project.

Matchbook Foldable® By Dinah Zike

Step 1 Fold a sheet of paper almost in half and make the back edge about 1–2 cm longer than the front edge.

Step 2 Find the midpoint of the shorter flap.

Step 3 Open the paper and cut the short side along the midpoint making two tabs.

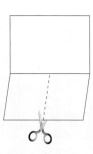

Step 4 Close the book and fold the tab over the short side.

Variations

A Make a single-tab matchbook by skipping Steps 2 and 3.

B Make two smaller matchbooks by cutting the single-tab matchbook in half.

Shutterfold Foldable® By Dinah Zike

Step 1 Begin as if you were folding a vertical sheet of paper in half, but instead of creasing the paper, pinch it to show the midpoint.

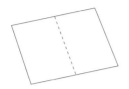

Step 2 Fold the top and bottom to the middle and crease the folds.

Variations

A Use the shutterfold on the horizontal axis.

B Create a center tab by leaving .5–2 cm between the flaps in Step 2.

PROJECT FORMAT
Use 11" × 17" or 12" × 18" paper and fold it to make a large shutterfold project.

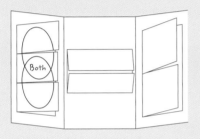

Four-Door Foldable® By Dinah Zike

Step 1 Make a shutterfold (see above).

Step 2 Fold the sheet of paper in half.

Step 3 Open the last fold and cut along the inside fold lines to make four tabs.

Variations

A Use the four-door book on the opposite axis.

B Create a center tab by leaving .5–2 cm between the flaps in Step 1.

Foldables Handbook • SR-35

Bound Book Foldable® By Dinah Zike

Step 1 Fold three sheets of paper in half. Place the papers in a stack, leaving about .5 cm between each top fold. Mark all three sheets about 3 cm from the outer edges.

Step 2 Using two of the sheets, cut from the outer edges to the marked spots on each side. On the other sheet, cut between the marked spots.

Step 3 Take the two sheets from Step 1 and slide them through the cut in the third sheet to make a 12-page book.

Step 4 Fold the bound pages in half to form a book.

Variation

A Use two sheets of paper to make an eight-page book, or increase the number of pages by using more than three sheets.

PROJECT FORMAT
Use two or more sheets of 11" × 17" or 12" × 18" paper and fold it to make a large bound book project.

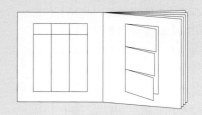

Accordian Foldable® By Dinah Zike

Step 1 Fold the selected paper in half vertically, like a *hamburger*.

Step 2 Cut each sheet of folded paper in half along the fold lines.

Step 3 Fold each half-sheet almost in half, leaving a 2 cm tab at the top.

Step 4 Fold the top tab over the short side, then fold it in the opposite direction.

Variations

A Glue the straight edge of one paper inside the tab of another sheet. Leave a tab at the end of the book to add more pages.

B Tape the straight edge of one paper to the tab of another sheet, or just tape the straight edges of nonfolded paper end to end to make an accordian.

C Use whole sheets of paper to make a large accordian.

SR-36 • Foldables Handbook

Layered Foldable® By Dinah Zike

Step 1 Stack two sheets of paper about 1–2 cm apart. Keep the right and left edges even.

Step 2 Fold up the bottom edges to form four tabs. Crease the fold to hold the tabs in place.

Step 3 Staple along the folded edge, or open and glue the papers together at the fold line.

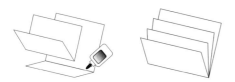

Variations

A Rotate the book so the fold is at the top or to the side.

B Extend the book by using more than two sheets of paper.

• •

Envelope Foldable® By Dinah Zike

Step 1 Fold a sheet of paper into a *taco*. Cut off the tab at the top.

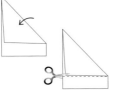

Step 2 Open the *taco* and fold it the opposite way making another *taco* and an X-fold pattern on the sheet of paper.

Step 3 Cut a map, illustration, or diagram to fit the inside of the envelope.

Step 4 Use the outside tabs for labels and inside tabs for writing information.

Variations

A Use 11″ × 17″ or 12″ × 18″ paper to make a large envelope.

B Cut off the points of the four tabs to make a window in the middle of the book.

Sentence Strip Foldable® By Dinah Zike

Step 1 Fold two sheets of paper in half vertically, like a *hamburger*.

Step 2 Unfold and cut along fold lines making four half sheets.

Step 3 Fold each half sheet in half horizontally, like a *hot dog*.

Step 4 Stack folded horizontal sheets evenly and staple together on the left side.

Step 5 Open the top flap of the first sentence strip and make a cut about 2 cm from the stapled edge to the fold line. This forms a flap that can be raisied and lowered. Repeat this step for each sentence strip.

Variations

A Expand this book by using more than two sheets of paper.

B Use whole sheets of paper to make large books.

Pyramid Foldable® By Dinah Zike

Step 1 Fold a sheet of paper into a *taco*. Crease the fold line, but do not cut it off.

Step 2 Open the folded sheet and refold it like a *taco* in the opposite direction to create an X-fold pattern.

Step 3 Cut one fold line as shown, stopping at the center of the X-fold to make a flap.

Step 4 Outline the fold lines of the X-fold. Label the three front sections and use the inside spaces for notes. Use the tab for the title.

Step 5 Glue the tab into a project book or notebook. Use the space under the pyramid for other information.

Step 6 To display the pyramid, fold the flap under and secure with a paper clip, if needed.

Single-Pocket or One-Pocket Foldable® By Dinah Zike

Step 1 Using a large piece of paper on a vertical axis, fold the bottom edge of the paper upwards, about 5 cm.

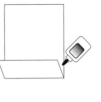

Step 2 Glue or staple the outer edges to make a large pocket.

> **PROJECT FORMAT**
> Use 11" × 17" or 12" × 18" paper and fold it vertically or horizontally to make a large pocket project.

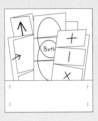

Variations

A Make the one-pocket project using the paper on the horizontal axis.

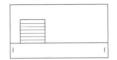

B To store materials securely inside, fold the top of the paper almost to the center, leaving about 2–4 cm between the paper edges. Slip the Foldables through the opening and under the top and bottom pockets.

Multi-Tab Foldable® By Dinah Zike

Step 1 Fold a sheet of notebook paper in half like a *hot dog*.

Step 2 Open the paper and on one side cut every third line. This makes ten tabs on wide ruled notebook paper and twelve tabs on college ruled.

Step 3 Label the tabs on the front side and use the inside space for definitions or other information.

Variation

A Make a tab for a title by folding the paper so the holes remain uncovered. This allows the notebook Foldable to be stored in a three-hole binder.

Reference Handbook

PERIODIC TABLE OF THE ELEMENTS

Legend:
- Element — Hydrogen
- Atomic number — 1
- Symbol — H
- Atomic mass — 1.01
- State of matter

- 🎈 Gas
- 💧 Liquid
- ⬛ Solid
- ⊙ Synthetic

A column in the periodic table is called a **group**.

A row in the periodic table is called a **period**.

	1	2	3	4	5	6	7	8	9
1	Hydrogen 1 H 1.01								
2	Lithium 3 Li 6.94	Beryllium 4 Be 9.01							
3	Sodium 11 Na 22.99	Magnesium 12 Mg 24.31							
4	Potassium 19 K 39.10	Calcium 20 Ca 40.08	Scandium 21 Sc 44.96	Titanium 22 Ti 47.87	Vanadium 23 V 50.94	Chromium 24 Cr 52.00	Manganese 25 Mn 54.94	Iron 26 Fe 55.85	Cobalt 27 Co 58.93
5	Rubidium 37 Rb 85.47	Strontium 38 Sr 87.62	Yttrium 39 Y 88.91	Zirconium 40 Zr 91.22	Niobium 41 Nb 92.91	Molybdenum 42 Mo 95.96	Technetium 43 Tc (98)	Ruthenium 44 Ru 101.07	Rhodium 45 Rh 102.91
6	Cesium 55 Cs 132.91	Barium 56 Ba 137.33	Lanthanum 57 La 138.91	Hafnium 72 Hf 178.49	Tantalum 73 Ta 180.95	Tungsten 74 W 183.84	Rhenium 75 Re 186.21	Osmium 76 Os 190.23	Iridium 77 Ir 192.22
7	Francium 87 Fr (223)	Radium 88 Ra (226)	Actinium 89 Ac (227)	Rutherfordium 104 Rf (267)	Dubnium 105 Db (268)	Seaborgium 106 Sg (271)	Bohrium 107 Bh (272)	Hassium 108 Hs (270)	Meitnerium 109 Mt (276)

The number in parentheses is the mass number of the longest lived isotope for that element.

Lanthanide series

Cerium 58 Ce 140.12	Praseodymium 59 Pr 140.91	Neodymium 60 Nd 144.24	Promethium 61 Pm (145)	Samarium 62 Sm 150.36	Europium 63 Eu 151.96

Actinide series

Thorium 90 Th 232.04	Protactinium 91 Pa 231.04	Uranium 92 U 238.03	Neptunium 93 Np (237)	Plutonium 94 Pu (244)	Americium 95 Am (243)

Periodic Table of the Elements

Legend:
- Metal
- Metalloid
- Nonmetal
- Recently discovered

	13	14	15	16	17	18
						Helium 2 He 4.00
	Boron 5 B 10.81	Carbon 6 C 12.01	Nitrogen 7 N 14.01	Oxygen 8 O 16.00	Fluorine 9 F 19.00	Neon 10 Ne 20.18

10	11	12						
			Aluminum 13 Al 26.98	Silicon 14 Si 28.09	Phosphorus 15 P 30.97	Sulfur 16 S 32.07	Chlorine 17 Cl 35.45	Argon 18 Ar 39.95
Nickel 28 Ni 58.69	Copper 29 Cu 63.55	Zinc 30 Zn 65.38	Gallium 31 Ga 69.72	Germanium 32 Ge 72.64	Arsenic 33 As 74.92	Selenium 34 Se 78.96	Bromine 35 Br 79.90	Krypton 36 Kr 83.80
Palladium 46 Pd 106.42	Silver 47 Ag 107.87	Cadmium 48 Cd 112.41	Indium 49 In 114.82	Tin 50 Sn 118.71	Antimony 51 Sb 121.76	Tellurium 52 Te 127.60	Iodine 53 I 126.90	Xenon 54 Xe 131.29
Platinum 78 Pt 195.08	Gold 79 Au 196.97	Mercury 80 Hg 200.59	Thallium 81 Tl 204.38	Lead 82 Pb 207.20	Bismuth 83 Bi 208.98	Polonium 84 Po (209)	Astatine 85 At (210)	Radon 86 Rn (222)
Darmstadtium 110 Ds (281)	Roentgenium 111 Rg (280)	Copernicium 112 Cn (285)	* Ununtrium 113 Uut (284)	* Ununquadium 114 Uuq (289)	* Ununpentium 115 Uup (288)	* Ununhexium 116 Uuh (293)		* Ununoctium 118 Uuo (294)

* The names and symbols for elements 113-116 and 118 are temporary. Final names will be selected when the elements' discoveries are verified.

Gadolinium 64 Gd 157.25	Terbium 65 Tb 158.93	Dysprosium 66 Dy 162.50	Holmium 67 Ho 164.93	Erbium 68 Er 167.26	Thulium 69 Tm 168.93	Ytterbium 70 Yb 173.05	Lutetium 71 Lu 174.97
Curium 96 Cm (247)	Berkelium 97 Bk (247)	Californium 98 Cf (251)	Einsteinium 99 Es (252)	Fermium 100 Fm (257)	Mendelevium 101 Md (258)	Nobelium 102 No (259)	Lawrencium 103 Lr (262)

Topographic Map Symbols

Topographic Map Symbols

Symbol	Description	Symbol	Description
————	Primary highway, hard surface	⌒	Index contour
══════	Secondary highway, hard surface	....	Supplementary contour
═══════	Light-duty road, hard or improved surface	⌢	Intermediate contour
=======	Unimproved road	⌾	Depression contours
+—+—+	Railroad: single track		
++==++	Railroad: multiple track	— — —	Boundaries: national
+++++	Railroads in juxtaposition	— — —	State
		— — - -	County, parish, municipal
▪▟▦	Buildings	— — - -	Civil township, precinct, town, barrio
♁ [†] cem	Schools, church, and cemetery	— · — · —	Incorporated city, village, town, hamlet
▫◻▨	Buildings (barn, warehouse, etc.)	· — · · —	Reservation, national or state
○ ○	Wells other than water (labeled as to type)	- - - - -	Small park, cemetery, airport, etc.
•••⊘	Tanks: oil, water, etc. (labeled only if water)	— ·· — ··	Land grant
○ ⚐	Located or landmark object; windmill	————	Township or range line, U.S. land survey
✕ ✕	Open pit, mine, or quarry; prospect	- - - - -	Township or range line, approximate location
	Marsh (swamp)		
	Wooded marsh	⌇⌇	Perennial streams
	Woods or brushwood	→—←	Elevated aqueduct
	Vineyard	○ ∼	Water well and spring
	Land subject to controlled inundation	⌇	Small rapids
	Submerged marsh	⌇	Large rapids
	Mangrove	▨	Intermittent lake
	Orchard	⌇	Intermittent stream
	Scrub	→==←	Aqueduct tunnel
	Urban area	▨	Glacier
		⌇	Small falls
x7369	Spot elevation	▨	Large falls
670	Water elevation	▨	Dry lake bed

Rocks

Rocks

Rock Type	Rock Name	Characteristics
Igneous (intrusive)	Granite	Large mineral grains of quartz, feldspar, hornblende, and mica. Usually light in color.
	Diorite	Large mineral grains of feldspar, hornblende, and mica. Less quartz than granite. Intermediate in color.
	Gabbro	Large mineral grains of feldspar, augite, and olivine. No quartz. Dark in color.
Igneous (extrusive)	Rhyolite	Small mineral grains of quartz, feldspar, hornblende, and mica, or no visible grains. Light in color.
	Andesite	Small mineral grains of feldspar, hornblende, and mica or no visible grains. Intermediate in color.
	Basalt	Small mineral grains of feldspar, augite, and possibly olivine or no visible grains. No quartz. Dark in color.
	Obsidian	Glassy texture. No visible grains. Volcanic glass. Fracture looks like broken glass.
	Pumice	Frothy texture. Floats in water. Usually light in color.
Sedimentary (detrital)	Conglomerate	Coarse grained. Gravel or pebble-size grains.
	Sandstone	Sand-sized grains 1/16 to 2 mm.
	Siltstone	Grains are smaller than sand but larger than clay.
	Shale	Smallest grains. Often dark in color. Usually platy.
Sedimentary (chemical or organic)	Limestone	Major mineral is calcite. Usually forms in oceans and lakes. Often contains fossils.
	Coal	Forms in swampy areas. Compacted layers of organic material, mainly plant remains.
Sedimentary (chemical)	Rock Salt	Commonly forms by the evaporation of seawater.
Metamorphic (foliated)	Gneiss	Banding due to alternate layers of different minerals, of different colors. Parent rock often is granite.
	Schist	Parallel arrangement of sheetlike minerals, mainly micas. Forms from different parent rocks.
	Phyllite	Shiny or silky appearance. May look wrinkled. Common parent rocks are shale and slate.
	Slate	Harder, denser, and shinier than shale. Common parent rock is shale.
Metamorphic (nonfoliated)	Marble	Calcite or dolomite. Common parent rock is limestone.
	Soapstone	Mainly of talc. Soft with greasy feel.
	Quartzite	Hard with interlocking quartz crystals. Common parent rock is sandstone.

Minerals

Minerals

Mineral (formula)	Color	Streak	Hardness Pattern	Breakage Properties	Uses and Other
Graphite (C)	black to gray	black to gray	1–1.5	basal cleavage (scales)	pencil lead, lubricants for locks, rods to control some small nuclear reactions, battery poles
Galena (PbS)	gray	gray to black	2.5	cubic cleavage perfect	source of lead, used for pipes, shields for X rays, fishing equipment sinkers
Hematite (Fe_2O_3)	black or reddish-brown	reddish-brown	5.5–6.5	irregular fracture	source of iron; converted to pig iron, made into steel
Magnetite (Fe_3O_4)	black	black	6	conchoidal fracture	source of iron, attracts a magnet
Pyrite (FeS_2)	light, brassy, yellow	greenish-black	6–6.5	uneven fracture	fool's gold
Talc ($Mg_3Si_4O_{10}(OH)_2$)	white, greenish	white	1	cleavage in one direction	used for talcum powder, sculptures, paper, and tabletops
Gypsum ($CaSO_4 \cdot 2H_2O$)	colorless, gray, white, brown	white	2	basal cleavage	used in plaster of paris and dry wall for building construction
Sphalerite (ZnS)	brown, reddish-brown, greenish	light to dark brown	3.5–4	cleavage in six directions	main ore of zinc; used in paints, dyes, and medicine
Muscovite ($KAl_3Si_3O_{10}(OH)_2$)	white, light gray, yellow, rose, green	colorless	2–2.5	basal cleavage	occurs in large, flexible plates; used as an insulator in electrical equipment, lubricant
Biotite ($K(Mg,Fe)_3(AlSi_3O_{10})(OH)_2$)	black to dark brown	colorless	2.5–3	basal cleavage	occurs in large, flexible plates
Halite (NaCl)	colorless, red, white, blue	colorless	2.5	cubic cleavage	salt; soluble in water; a preservative

Minerals

Minerals

Mineral (formula)	Color	Streak	Hardness	Breakage Pattern	Uses and Other Properties
Calcite ($CaCO_3$)	colorless, white, pale blue	colorless, white	3	cleavage in three directions	fizzes when HCl is added; used in cements and other building materials
Dolomite ($CaMg(CO_3)_2$)	colorless, white, pink, green, gray, black	white	3.5–4	cleavage in three directions	concrete and cement; used as an ornamental building stone
Fluorite (CaF_2)	colorless, white, blue, green, red, yellow, purple	colorless	4	cleavage in four directions	used in the manufacture of optical equipment; glows under ultraviolet light
Hornblende ($(CaNa)_{2-3}(Mg,Al,Fe)_5-(Al,Si)_2Si_6O_{22}(OH)_2$)	green to black	gray to white	5–6	cleavage in two directions	will transmit light on thin edges; 6-sided cross section
Feldspar ($KAlSi_3O_8$), ($NaAlSi_3O_8$), ($CaAl_2Si_2O_8$)	colorless, white to gray, green	colorless	6	two cleavage planes meet at 90° angle	used in the manufacture of ceramics
Augite ($(Ca,Na)(Mg,Fe,Al)(Al,Si)_2O_6$)	black	colorless	6	cleavage in two directions	square or 8-sided cross section
Olivine ($(Mg,Fe)_2SiO_4$)	olive, green	none	6.5–7	conchoidal fracture	gemstones, refractory sand
Quartz (SiO_2)	colorless, various colors	none	7	conchoidal fracture	used in glass manufacture, electronic equipment, radios, computers, watches, gemstones

Weather Map Symbols

Sample Station Model

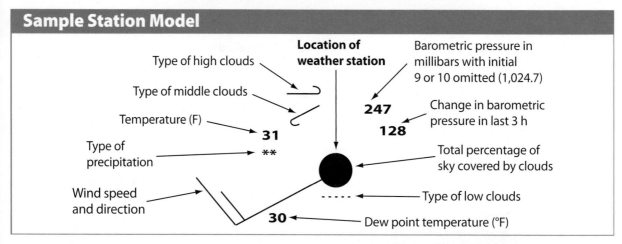

Sample Plotted Report at Each Station

Precipitation		Wind Speed and Direction		Sky Coverage		Some Types of High Clouds	
≡	Fog	○	0 calm	○	No cover	⌒	Scattered cirrus
★	Snow	/	1–2 knots	⊖	1/10 or less	⌒⊃	Dense cirrus in patches
●	Rain	⌄	3–7 knots	◔	2/10 to 3/10	⌒c	Veil of cirrus covering entire sky
⊼	Thunderstorm	⌄	8–12 knots	◓	4/10	⌒c	Cirrus not covering entire sky
,	Drizzle	⌄	13–17 knots	◐	–		
∇	Showers	⌄	18–22 knots	◕	6/10		
		⌄	23–27 knots	◕	7/10		
		⌄	48–52 knots	⦶	Overcast with openings		
			1 knot = 1.852 km/h	●	Completely overcast		

Some Types of Middle Clouds		Some Types of Low Clouds		Fronts and Pressure Systems	
∠	Thin altostratus layer	⌒	Cumulus of fair weather	Ⓗ or High Ⓛ or Low	Center of high- or low-pressure system
∠	Thick altostratus layer	∪	Stratocumulus	▲▲▲▲	Cold front
⌒	Thin altostratus in patches	-----	Fractocumulus of bad weather	⌒⌒⌒⌒	Warm front
⌒	Thin altostratus in bands	—	Stratus of fair weather	▲⌒▲⌒	Occluded front
				⌒▲⌒▲	Stationary front

SR-46 • Reference Handbook

Glossary/Glosario

Multilingual eGlossary

A science multilingual glossary is available on the science Web site. The glossary includes the following languages.

Arabic
Bengali
Chinese
English
Haitian Creole
Hmong
Korean
Portuguese
Russian
Spanish
Tagalog
Urdu
Vietnamese

Cómo usar el glosario en español:
1. Busca el término en inglés que desees encontrar.
2. El término en español, junto con la definición, se encuentran en la columna de la derecha.

Pronunciation Key

Use the following key to help you sound out words in the glossary.

a	back (BAK)	ew	food (FEWD)
ay	day (DAY)	yoo	pure (PYOOR)
ah	father (FAH thur)	yew	few (FYEW)
ow	flower (FLOW ur)	uh	comma (CAH muh)
ar	car (CAR)	u (+ con)	rub (RUB)
e	less (LES)	sh	shelf (SHELF)
ee	leaf (LEEF)	ch	nature (NAY chur)
ih	trip (TRIHP)	g	gift (GIHFT)
i (i + com + e)	idea (i DEE uh)	j	gem (JEM)
oh	go (GOH)	ing	sing (SING)
aw	soft (SAWFT)	zh	vision (VIH zhun)
or	orbit (OR buht)	k	cake (KAYK)
oy	coin (COYN)	s	seed, cent (SEED, SENT)
oo	foot (FOOT)	z	zone, raise (ZOHN, RAYZ)

English — **A** — **Español**

acid precipitation/atmosphere (AT muh sfihr) — **precipitación ácida/atmósfera**

acid precipitation: precipitation that has a lower pH than that of normal rainwater (pH 5.6). (p. 435)

air mass: a large area of air that has uniform temperature, humidity, and pressure. (p. 460)

air pollution: the contamination of air by harmful substances including gases and smoke. (p. 434)

air pressure: the pressure that a column of air exerts on the air, or a surface, below it. (p. 452)

atmosphere (AT muh sfihr): a thin layer of gases surrounding Earth. (p. 409)

precipitación ácida: precipitación que tiene un pH más bajo que el del agua de la lluvia normal (pH 5.6). (pág. 435)

masa de aire: gran área de aire que tiene temperatura, humedad y presión uniformes. (pág. 460)

polución del aire: contaminación del aire por sustancias dañinas, como gases y humo. (pág. 434)

presión del aire: presión que una columna de aire ejerce sobre el aire o sobre la superficie debajo de ella. (pág. 452)

atmósfera: capa delgada de gases que rodean la Tierra. (pág. 409)

blizzard/El Nino/Southern Oscillation   **ventisca/El Niño/Oscilación meridional**

B

blizzard: a violent winter storm characterized by freezing temperatures, strong winds, and blowing snow. (p.467)

ventisca: tormenta violenta de invierno caracterizada por temperaturas heladas, vientos fuertes, y nieve que sopla. (pág. 467)

C

climate: the long-term average weather conditions that occur in a particular region. (p. 487)

computer model: detailed computer programs that solve a set of complex mathematical formulas. (p. 474)

conduction (kuhn DUK shun): the transfer of thermal energy due to collisions between particles. (p. 421)

convection: the circulation of particles within a material caused by differences in thermal energy and density. (p. 421)

clima: promedio a largo plazo de las condiciones del tiempo atmosférico de una región en particular. (pág. 487)

modelo de computadora: programas de computadora que resuelven un conjunto de fórmulas matemáticas complejas. (pág. 474)

conducción: transferencia de energía térmica debido a colisiones entre partículas. (pág. 421)

convección: circulación de partículas en el interior de un material causada por diferencias en la energía térmica y la densidad. (pág. 421)

D

deforestation: the removal of large areas of forests for human purposes. (p. 507)

dew point: temperature at which air is fully saturated because of decreasing temperatures while holding the amount of moisture constant. (p. 453)

Doppler radar: a specialized type of radar that can detect precipitation as well as the movement of small particles, which can be used to approximate wind speed. (p. 472)

drought: a period of below-average precipitation. (p.501)

deforestación: eliminación de grandes áreas de bosques con propósitos humanos. (pág. 507)

punto de rocío: temperatura en la cual el aire está completamente saturado debido a la disminución en las temperaturas aunque mantiene constante la cantidad de humedad. (pág. 453)

radar Dopler: tipo de radar especializado que detecta tanto la precipitación como el movimiento de partículas pequeñas, que se pueden usar para determinar la velocidad aproximada del viento. (pág. 472)

sequía: período de bajo promedio de precipitación. (pág. 501)

E

El Nino/Southern Oscillation: the combined ocean and atmospheric cycle that results in weakened trade winds across the Pacific Ocean. (p. 500)

El Niño/Oscilación meridional: ciclo atmosférico y oceánico combinado que produce el debilitamiento de los vientos alisios en el Océano Pacífico. (pág. 500)

F

front: a boundary between two air masses. (p. 462)

frente: límite entre dos masas de aire. (pág. 462)

G

global climate model: a set of complex equations used to predict future climates. (p. 509)

global warming: an increase in the average temperature of Earth's surface. (p. 506)

greenhouse gas: a gas in the atmosphere that absorbs Earth's outgoing infrared radiation. (p. 506)

modelo de clima global: conjunto de ecuaciones complejas para predecir climas futuros. (pág. 509)

calentamiento global: incremento en la temperatura promedio de la superficie de la Tierra. (pág. 506)

gas de invernadero: gas en la atmósfera que absorbe la salida de radiación infrarroja de la Tierra. (pág. 506)

H

high-pressure system: a large body of circulating air with high pressure at its center and lower pressure outside of the system. (p. 459)
humidity (hyew MIH duh tee): the amount of water vapor in the air. (p. 452)
hurricane: an intense tropical storm with winds exceeding 119 km/h. (p. 466)

sistema de alta presión: gran cuerpo de aire circulante con presión alta en el centro y presión más baja fuera del sistema. (pág. 459)
humedad: cantidad de vapor de agua en el aire. (pág. 452)
huracán: tormenta tropical intensa con vientos que exceden los 119 km/h. (pág. 466)

I

ice age: a period of time when a large portion of Earth's surface is covered by glaciers. (p. 496)
interglacial: a warm period that occurs during an ice age. (p. 496)
ionosphere: a region within the mesosphere and thermosphere containing ions. (p. 413)
isobar: lines that connect all places on a map where pressure has the same value. (p. 473)

era del hielo: período de tiempo cuando los glaciares cubren una gran porción de la superficie de la Tierra. (pág. 496)
interglacial: período tibio que ocurre durante una era del hielo. (pág. 496)
ionosfera: región entre la mesosfera y la termosfera que contiene iones. (pág. 413)
isobara: línea que conectan todos los lugares en un mapa donde la presión tiene el mismo valor. (pág. 473)

J

jet stream: a narrow band of high winds located near the top of the troposphere. (p. 429)

corriente de chorro: banda angosta de vientos fuertes cerca de la parte superior de la troposfera. (pág. 429)

land breeze/relative humidity   **brisa terrestre/humedad relativa**

L

land breeze: a wind that blows from the land to the sea due to local temperature and pressure differences. (p. 436)

low-pressure system: a large body of circulating air with low pressure at its center and higher pressure outside of the system. (p. 459)

brisa terrestre: viento que sopla desde la tierra hacia el mar debido a diferencias en la temperatura local y la presión. (pág. 436)

sistema baja presión: gran cuerpo de aire circulante con presión baja en el centro y presión más alta fuera del sistema. (pág. 459)

M

microclimate: a localized climate that is different from the climate of the larger area surrounding it. (p. 491)

monsoon: a wind circulation pattern that changes direction with the seasons. (p. 501)

microclima: clima localizado que es diferente del clima de área más extensa que lo rodea. (pág. 491)

monsón: patrón de viento circulante que cambia de dirección con las estaciones. (pág. 501)

O

ozone layer: the area of the stratosphere with a high concentration of ozone. (p. 412)

capa de ozono: área de la estratosfera con gran concentración de ozono. (pág. 412)

P

particulate (par TIH kyuh lut) matter: the mix of both solid and liquid particles in the air. (p. 430)

photochemical smog: air pollution that forms from the interaction between chemicals in the air and sunlight. (p. 435)

polar easterlies: cold winds that blow from the east to the west near the North Pole and South Pole. (p. 429)

precipitation: water, in liquid or solid form, that falls from the atmosphere. (p. 455)

partículas en suspensión: mezcla de partículas tanto sólidas como líquidas en el aire. (pág. 430)

smog fotoquímico: polución del aire que se forma de la interacción entre los químicos en el aire y la luz solar. (pág. 435)

brisas polares: vientos fríos que soplan del este al oeste cerca del Polo Norte y del Polo Sur. (pág. 429)

precipitación: agua, de forma líquida o sólida, que cae de la atmósfera. (pág. 455)

R

radiation: the transfer of thermal energy by electromagnetic waves. (p. 418)

rain shadow: an area of low rainfall on the downwind slope of a mountain. (p. 489)

relative humidity: the amount of water vapor present in the air compared to the maximum amount of water vapor the air could contain at that temperature. (p. 453)

radiación: transferencia de energía térmica mediante ondas electromagnéticas. (pág. 418)

sombra de lluvia: área de baja precipitación en la ladera de sotavento de una montaña. (pág. 489)

humedad relativa: cantidad de vapor de agua presente en el aire comparada con la cantidad máxima de vapor de agua que el aire podría contener en esa temperatura. (pág. 453)

sea breeze/weather **brisa marina/tiempo atmosférico**

S

sea breeze: a wind that blows from the sea to the land due to local temperature and pressure differences. (p. 430)
specific heat: the amount of thermal energy (joules) needed to raise the temperature of 1 kg of material 1°C. (p. 489)
stability: whether circulating air motions will be strong or weak. (p. 422)
stratosphere (STRA tuh sfihr): the atmospheric layer directly above the troposphere. (p. 412)
surface report: a description of a set of weather measurements made on Earth's surface. (p. 471)

brisa marina: viento que sopla del mar hacia la tierra debido a diferencias en la temperatura local y la presión. (pág. 430)
calor específico: cantidad de energía térmica (julios) requerida para subir la temperatura de 1 kg de materia a 1°C. (pág. 489)
estabilidad: condición en la que los movimientos del aire circulante pueden ser fuertes o débiles. (pág. 422)
estratosfera: capa atmosférica justo arriba de la troposfera. (pág. 412)
informe de superficie: descripción de un conjunto de mediciones del tiempo realizadas en la superficie de la Tierra. (pág. 471)

T

temperature inversion: a temperature increase as altitude increases in the troposphere. (p. 423)
tornado: a violent, whirling column of air in contact with the ground. (p. 465)
trade winds: steady winds that flow from east to west between 30°N latitude and 30°S latitude. (p. 429)
troposphere (TRO puh sfihr): the atmospheric layer closest to Earth's surface. (p. 412)

inversión de temperatura: aumento de la temperatura en la troposfera a medida que aumenta la altitud. (pág. 423)
tornado: columna de aire violenta y rotativa en contacto con el suelo. (pág. 465)
vientos alisios: vientos constantes que soplan del este al oeste entre 30°N de latitud y 30°S de latitud. (pág. 429)
troposfera: capa atmosférica más cercana a la Tierra. (pág. 412)

U

upper-air report: a description of wind, temperature, and humidity conditions above Earth's surface. (p. 471)

informe del aire superior: descripción de las condiciones del viento, de la temperatura y de la humedad por encima de la superficie de la Tierra. (pág. 471)

W

water cycle: the series of natural processes by which water continually moves throughout the hydrosphere. (p. 455)
water vapor: water in its gaseous form. (p. 410)
weather: the atmospheric conditions, along with short-term changes, of a certain place at a certain time. (p. 451)

ciclo del agua: serie de procesos naturales por los que el cual el agua se mueve continuamente en toda la hidrosfera. (pág. 455)
vapor de agua: agua en forma gaseosa. (pág. 410)
tiempo atmosférico: condiciones atmosféricas, junto con cambios a corto plazo, de un lugar determinado a una hora determinada. (pág. 451)

westerlies: steady winds that flow from west to east between latitudes 30°N and 60°N, and 30°S and 60°S. (p. 429)

wind: the movement of air from areas of high pressure to areas of low pressure. (p. 427)

vientos del oeste: vientos constantes que soplan de oeste a este entre latitudes 30°N y 60°N, y 30°S y 60°S. (pág. 429)

viento: movimiento del aire desde áreas de alta presión hasta áreas de baja presión. (pág. 427)

Index

Absorption   *Italic numbers* = illustration/photo   **Bold numbers** = vocabulary term   Coriolis effect
lab = indicates entry is used in a lab on this page

A

Absorption
explanation of, 419
Academic Vocabulary, 418, 464, 500. *See also* **Vocabulary**
Acid precipitation
effects of, **435**
explanation of, **435**
formation of, 436 *lab*
Acid rain
formation of, 436 *lab*
Aerosol(s)
explanation of, 507
Air circulation
explanation of, 422
global wind belts and, 428, *428*
three-cell model of circulation, 428, *428*, 432 *lab*
Air current(s)
global winds and, 427, *427–429*
local winds and, 430, *430*
Air mass(es)
Antarctic air masses, 460
Arctic air masses, 460, 461
classification of, *460*, 460–461
explanation of, **460**
Air pollution. *See also* **Pollution**
acid precipitation as, 435, 436 *lab*
explanation of, **436**
indoor, 438
monitoring of, 437
movement of, 436, *436*
particulate matter as, 436
smog as, 435, *435*
sources of, 436
temperature inversion and, 423
Air pressure
altitude and, 414, *414*
explanation of, 452, **452**
observation of, 461 *lab*
Air quality
monitoring of, 437
standards for, 437
trends in, 438, *438*
Air temperature
explanation of, 452
pressure and, 459, 459 *lab*
water vapor and, 453, *453*
Air. *See also* **Atmosphere**
movement of, 427 *lab*
stable, 423
unstable, 423
Albedo
explanation of, 494
Altitude
air pressure and, 414, *414*
temperature and, 414, *414*, 487, 488, *488*

Anemometer
explanation of, 452, *452*
Animal(s)
adaptations to climate by, 492
Antarctica
hole in ozone layer above, 416
temperature in, 487
Argon
in atmosphere, 411
Arizona
monsoon in, 501
Ash
in atmosphere, 411, *411*
Atmosphere. *See also* **Air**
air pressure and, 414, *414*
composition of, 411
explanation of, **409**
importance of, 409
layers of, *412*, 412–413
origins of, 410
solid particles in, 410 *lab*, 411
stable, 423
temperature and, 414
three-cell model of circulation in, 428, *428*, 432 *lab*
unstable, 423
Auroras
explanation of, 413, *413*

B

Barometer(s)
explanation of, *452*
Barometric pressure
explanation of, 452
Big Idea, 406, 442, 448, 478, 484, 513, 514
Review, 445, 481, 517
Bjerknes, Jacob, 462
Blizzard(s)
explanation of, 467, *467*
Bromine
in atmosphere, 416

C

Carbon dioxide
in atmosphere, 409–411
as greenhouse gas, 506, *506*
infrared radiation and, 420
sources of, 507
vehicle emissions of, 509 *lab*
Careers in Science, 416, 503
Chapter Review, 444–445, 480–481, 516–517
Cherrapunji, India, 501
Chlorine
in atmosphere, 416

Chlorofluorocarbons (CFCs)
ozone layer and, 416
Clear Air Act (1970), 437
Climate change. *See also* **Global warming**
environmental impact of, 508
human impact on, 506–507
methods to predict, 509, 510
methods to reduce, 510
ozone layer and, 416
regional and global, 505, *505*
sources of information on, 496, *496*, 503
Climate cycle(s)
causes of long-term, 497, *497*
explanation of, 496
ice ages and, 496, 497
short-term, 498–501
Climate. *See also* **Temperature; Weather**
adaptations to, 492, *492*
comparison of, 487 *lab*
effect of albedo on, 494
explanation of, **487**
factors affecting, 487–489, 505 *lab*
methods to classify, 490, *491*
microclimates, 490
Cloud(s)
cumulus, 464, *464*
with droplets, 507, *507*
explanation of, 454
formation of, 451 *lab*, 454, 455
variables used to describe, 451–455, *452*, *454*, *455*
water cycle and, 455
Coastline(s)
climate on, 487, 489
Cold front(s)
explanation of, 462, *462*
Cold wave(s)
explanation of, 501
Common Use. *See* **Science Use v. Common Use**
Condensation
in water cycle, 455
Conduction
explanation of, **421,** 425 *lab*, 440 *lab*
Continental climate
explanation of, *491*
vegetation in, 492
Continental polar air mass(es)
explanation of, 461
Continental tropical air mass(es)
explanation of, 461
Convection cell
explanation of, 428, *428*
Convection
explanation of, **421,** 422, 425 *lab*
Coriolis effect
explanation of, 428, 429

I-2 • Index

C

Critical thinking, 415, 424, 431, 439, 456, 468, 475, 481, 493, 502, 511, 517
Cumulus cloud(s)
explanation of, 464

D

Deciduous tree(s), 492
Deforestation
carbon dioxide and, 507
explanation of, 506, **507**
Dew point
air temperature and, 453 *lab*
explanation of, **453**
Doldrums
explanation of, 429
Dominate
explanation of, **464**
Doppler radar
explanation of, 472
Downdraft(s)
explanation of, 464
Drought
effects of, 508
explanation of, 501
Dry climate
adaptations to, 492, *492*
explanation of, *491*

E

Earth
curved surface of, 488
effect of orbit and tilt of axis on climate of, 496, 496 *lab*, 497, *497*, 498, *498*
energy on, 419
revolution of, *499*, 499
El Niño/Southern Oscillation (ENSO)
explanation of, **500**
Electromagnetic wave(s)
transfer of energy by, 418
Energy
on Earth, 419
from Sun, 418, 427, 440 *lab*
Environmental impact
of climate change, 508
Environmental protection
technological trends for, 510, *510*
Equator
solar energy and, 488
temperature in, 487
Equinox
explanation of, 499, *499*
Evaporation
in water cycle, 455
Exosphere
explanation of, *412*, 413
temperature in, 414, *414*

F

Fog
explanation of, 454
Foldables, 413, 422, 428, 435, 443, 454, 460, 472, 479, 490, 498, 505, 515

Fossil fuel(s)
carbon dioxide released from burning, 507
Front(s)
cold, 462, *462*
explanation of, **462**
occluded, 463, *463*
stationary, 463, *463*
warm, *462*, 463
Frostbite, 467
Fujita, Ted, 465

G

Gas(es)
in atmosphere, 411
Geologist(s)
explanation of, 503
Glaciers
explanation of, 496
Global climate model (GCM)
explanation of, 509
Global warming. *See also* **Climate change**
explanation of, **506**
hurricanes and, 457
ozone layer and, 416
Gravity
explanation of, 414
Green building
explanation of, 510, *510*
Greenhouse effect. *See also* **Climate change**
explanation of, 420, *420*, 506
modeling of, 512–513 *lab*
Greenhouse gas(es). *See also* **Climate change**
explanation of, **506**
methods to reduce, 510, *510*
sources of, 507
Gulf Stream
explanation of, 489

H

Hail
explanation of, 455 (See also
Haze
explanation of, 436
Heat wave(s)
drought and, 501
effects of, 508
Heat. *See also* **Thermal energy**
explanation of, 418
latent, 421
High-pressure system
air masses and, 460
explanation of, *459*, **459**
Holocene Epoch
explanation of, 497
Human population growth
projections for, 510, *510*
Humans
impact on climate change, 506–507
Humidity
explanation of, **452**
relative, 453

Hurricane Katrina
global warming and, 457
Hurricane(s)
explanation of, **466**
formation of, 466, *466*
global warming and, 457
Hypothermia, 467

I

Ice age(s)
description of past, 497
explanation of, **496**
Ice core(s)
analysis of, 503
explanation of, 496, *496*
method to collect, 503
Ice sheet(s)
explanation of, 496
in most recent ice age, 497
Ice storm(s)
as driving hazard, 467
Indoor air pollution. *See also* **Air pollution**
explanation of, 438
Infrared radiation (IR)
explanation of, 418
greenhouse effect and, 420
Infrared satellite image(s)
explanation of, 472, *472*
Interglacial(s)
explanation of, **496,** 497
Intergovernmental Panel on Climate Change (IPCC), 506
Interpret Graphics, 415, 424, 431, 439, 456, 468, 475, 493, 502, 511
Ionosphere
explanation of, *413*, **413**
Isobar(s)
explanation of, **473**
Isotherm(s)
explanation of, 473

J

Jet stream(s)
explanation of, 429, *429*

K

Key Concepts, 408, 417, 426, 435, 450, 458, 470, 486, 495, 504
Check, 410, 411, 413, 414, 421, 422, 427, 428, 429, 436, 438, 451, 455, 459, 461, 463, 467, 472, 474, 487, 490, 497, 498, 501, 507, 509
Summary, 442, 478, 514
Understand, 415, 424, 431, 439, 456, 468, 475, 480, 493, 502, 511, 516
Kinetic energy
explanation of, **452**
Köppen, Wladimir, 490

L

La Niña
explanation of, 501 *lab*

Lab, 440–441, 476–477, 512–513. *See also* **Launch Lab; MiniLab; Skill Practice**
Land breeze
　explanation of, *430*, **430**
Latent heat
　explanation of, 421
Latitude
　temperature and, 487, 488, *488*, 498
Launch Lab, 409, 418, 427, 436, 451, 459, 471, 487, 496, 505
Lesson Review, 415, 424, 431, 439, 456, 468, 475, 493, 502, 511
Liquid
　explanation of, **410**
Little Ice Age
　explanation of, 497
Low-pressure system
　explanation of, *459*, **459**

M

Maritime polar air mass(es)
　explanation of, 461
Maritime tropical air mass(es)
　explanation of, 461
Math Skills, 438, 439, 445, 461, 468, 481, 508, 511, 517
Mesosphere
　explanation of, *412*, 413
　temperature in, 414, *414*
Meteorologist(s)
　explanation of, 451, 472, *473*
Methane
　as greenhouse gas, 506
　infrared radiation and, 420
Microclimate(s)
　explanation of, **490**
　location of, 492 *lab*
Mild climate
　explanation of, *491*
MiniLab, 410, 423, 429, 437, 453, 461, 474, 492, 501, 509. *See also* **Lab**
Monsoon
　explanation of, **501**
Motor vehicle(s)
　emissions of, 509 *lab*
　hybrid, 510
Mountain wave(s)
　explanation of, 422, *422*
Mountain(s)
　rain shadows in, 489, *489*
　weather in, 487

N

NASA
　ozone layer research by, 416
Nitrogen
　in atmosphere, 410, 411, *411*
Nitrous oxide
　in atmosphere, 411
Non-point source pollution
　explanation of, 436
North Atlantic Oscillation (NAO)
　explanation of, **500**

North Pole
　solar energy and, 488, *488*

O

Occluded front(s)
　explanation of, 463
Ocean current(s)
　climate and, 489
Ocean(s)
　warming of, 457
Oxygen
　in atmosphere, 409–411
Ozone layer
　explanation of, **412**
　hole in, 416
Ozone
　in atmosphere, 411, 414, 435
　explanation of, 412
　ground-level, 437, 438

P

Particulate matter
　explanation of, **436**
Permafrost
　explanation of, 492
Phenomenon
　explanation of, **500**
Photochemical smog
　explanation of, **435**
Photosynthesis
　explanation of, 410, 507
Point source pollution
　explanation of, 436
Polar climate
　explanation of, *491*
　vegetation in, 492
Polar regions
　temperature in, 488
Pollution. *See also* **Air pollution**
　acid rain as, 436 *lab*
　methods to reduce, 510, *510*
　non-point source, 436
　particulate, 436
　point-source, 436
　temperature inversion and, 423
Precipitation
　climate change and, 508
　explanation of, **454, 455, 489**
　types of, 455
　water cycle and, 455
Pressure system(s)
　explanation of, 459
Process
　explanation of, **418**

R

Radar
　Doppler, 472
　measurement of precipitation by, 472
Radiation
　absorption of, 419, *419*
　balance in, 420, *420*
　explanation of, **418**

Radio wave(s), 413, *413*
Rain shadow(s)
　explanation of, *489*, **489**
Rain. *See also* **Precipitation**
　adaptations to, 492
　explanation of, 455
　freezing, 467
Reading Check, 409, 412, 414, 418, 420, 423, 430, 437, 452, 453, 454, 462, 465, 473, 489, 492, 496, 499, 500, 505, 506, 509, 510
Recycling
　to control greenhouse gases and pollution, 510
Reflection
　of radiation, 419, *419*
Relative humidity
　explanation of, **453**
Renewable energy
　wind as, 427 *lab*
Review Vocabulary, 410, 451, 452, 489. *See also* **Vocabulary**
Revolution
　explanation of, **499**

S

Sahara
　temperature in, 487
Satellite(s)
　weather information from images taken by, 472, *472*
Science & Society, 457
Science Methods, 441, 477, 513
Science Use v. Common Use, 462, 419, 499. *See also* **Vocabulary**
Sea breeze
　explanation of, *430*, **430**
Season(s)
　explanation of, 498
　monsoon winds and, *496*, 501
Shindell, Drew, 416
Skill Practice, 425, 432, 469, 494. *See also* **Lab**
Sleet. *See also* **Precipitation**
　explanation of, 455
Smog
　explanation of, 435
Snow. *See also* **Precipitation**
　albedo of, 494
　as driving hazard, 467
　explanation of, 455
Solar energy
　climate cycles and, 497, 498, *498*
　reflected back into atmosphere, 494
　temperatures and, 487, 488, *488*
Solstice
　explanation of, 499, *499*
South Pole
　solar energy and, 488, *488*
Specific heat
　explanation of, **489**
Stability
　air, 422–423
　explanation of, **422**
Standardized Test Practice, 446–447, 482–483, 518–519

Station model
explanation of, 473, *473*, 474
Stationary front(s)
explanation of, 463
Stratosphere
explanation of, *412*, **412**
ozone in, 414, 416, 435
Study Guide, 442–443, 478–479, 514–515
Sulfur dioxide
in atmosphere, 411
Sun
energy from, 418, 438 *lab*
reflection of rays from, 494
Sunlight
unequal distribution of, 427
Surface report(s)
explanation of, **471**

T

Temperature inversion
explanation of, **423**
identification of, 423 *lab*
pollution and, 436, *436*
Temperature. *See also* **Climate; Weather**
adaptations to, 492
of air, 452, 453, *453*, 459 *lab*
altitude and, 414, *414*
effect of greenhouse gases on, 506
effects of increasing, 508–509
trends in, 505, *505*
variations in, 487, 488, *488*
Thermal energy. *See also* **Heat**
effects of, 492
explanation of, 418 *lab*
in ocean water, 489
production of, 418
transfer of, 421, *421*
Thermosphere
explanation of, *412*, 413
temperature in, 414, *414*
Thompson, Lonnie, 503
Thunderstorm(s)
explanation of, 423, 464, *464*, 465
safety precautions during, 467
Tornado Alley, 465
Tornado(es)
classification of, 465
explanation of, 465, *465*
formation of, 465
Trade wind(s)
explanation of, **429,** 500
Tropical climate
explanation of, *491*
Tropical cyclone(s), 466
Tropics
sunlight in, 427
Troposphere
explanation of, *412*, **412**
jet streams in, 429
temperature in, 414, *414*, 423
Typhoon(s), 466

U

U.S. National Weather Service, 467
Ultraviolet (UV) light
explanation of, 418
Updraft(s)
explanation of, 464, *465*
tornado formation and, 465
Upper-air report(s)
explanation of, **471**
Upwelling
explanation of, 500
Urban areas
ozone in, 437
population growth projections for, 510
temperatures in, 487, 490
Urban heat island
explanation of, 491, *491*

V

Variable(s)
explanation of, **451**
Vegetation
adaptations to climate by, 492
classifying climate by native, 490
in mountains, 489
Vehicle emissions
hybrid vehicles and, 510
variations in, 509 *lab*
Visible light satellite image(s)
explanation of, 472, *472*
Visual Check, 411, 412, 414, 419, 423, 428, 430, 436, 452, 455, 460, 462, 464, 466, 473, 488, 489, 490, 499, 500, 505
Vocabulary, 407, 408, 417, 426, 435, 442, 449, 450, 458, 470, 478, 485, 486, 495, 504, 511, 514. *See also* **Academic Vocabulary; Review Vocabulary; Science Use v. Common Use; Word Origin**
Use, 415, 424, 431, 439, 443, 456, 468, 475, 479, 493, 502, 515
Volcanic eruption(s)
solid particles in atmosphere from, 411, *411*

W

Warm front(s)
explanation of, *462*, 463
Water bodies
climatic effect of large, 487, 489–490
Water cycle
explanation of, *455*, **455**
Water vapor
air temperature and, 453, *453*
in atmosphere, 411
explanation of, **410**
as greenhouse gas, 506
infrared radiation and, 420
water cycle and, 455

Water
specific heat of, 489
Weather forecast(s)
explanation of, 471
methods for, 476–477 *lab*
satellite and radar images for, 472
station models use for, 473, 474 *lab*
surface reports as, 471
understanding, 471 *lab*
upper-air reports as, 471
use of technology in, 474, *474*
Weather map(s)
explanation of, 472–473
of temperature and pressure, 473, *473*
Weather. *See also* **Climate; Temperature**
cycles of, 496–501
explanation of, **451,** 487
reasons for change in, 469
safety precautions during severe, 467
types of severe, 464, 464–467, *465, 466, 467*
variables related to, 451–455, *452, 453, 454, 455*
water cycle and, 455, *455*
What do you think?, 407, 415, 424, 431, 439, 449, 456, 468, 475, 485, 493, 502, 511
Wind belts
explanation of, 427, *428*
global, 428–429
Wind(s)
air pollution and, 436
explanation of, 427, 452
global, *427*, 427–429
local, 430, *430*
measurement of, 452, *452*
monsoon, 501
pressure systems and, 459
as renewable energy source, 427 *lab*
trade, 429
Windchill
explanation of, 467
Winter storm(s)
explanation of, 467
Word Origin, 409, 421, 436, 454, 466, 473, 490, 497, 506. *See also* **Vocabulary**
Writing In Science, 445, 481, 517

Credits

Photo Credits

Front Cover blickwinkle/Alamy; **Spine-Back Cover** Walter Geiersperger/CORBIS; **Connect Ed** (t)Richard Hutchings, (c)Getty Images, (b)Jupiter Images/Thinkstock/Alamy; **i** Thinkstock/Getty Images; **iv** Ransom Studios **viii–ix** The McGraw-Hill Companies; **ix** (b)Fancy Photography/Veer **404** Reuters/Corbis; **405** (l)John W. van de Lindt/Colorado State University; (r) Daniel Cox/O.H. Hinsdale Research Laboratory/Oregon State University; **406–407** Daniel H. Bailey/Alamy; **408** CORBIS; **409,410** Hutchings Photography/Digital Light Source; **411** (t)PhotoLink/Getty Images; (b)C. Sherburne/PhotoLink/Getty Images; **413** Per Breiehagen/Getty Images; **414** CORBIS; **415** (t)PhotoLink/Getty Images; (c)Per Breiehagen/Getty Images; (r)Hutchings Photography/Digital Light Source; **416** (t)American Museum of Natural History; (b)Pedro Guzman; (bkgd)PhotoLink/Getty Images; **417** John King/Alamy; **418** Hutchings Photography/Digital Light Source; **419** Eric James/Alamy; **422** (t)C. Sherburne/PhotoLink/Getty Images; (b)James Brunker/Alamy; **424** Eric James/Alamy; **425** (2,8) Macmillan/McGraw-Hill; (others)Hutchings Photography/Digital Light Source; **426** Lester Lefkowitz/Getty Images; **427** Hutchings Photography/Digital Light Source; **429** (t)CORBIS; (b)Hutchings Photography/Digital Light Source; **431** Lester Lefkowitz/Getty Images; **432** (3,4)Macmillan/McGraw-Hill; (others)Hutchings Photography/Digital Light Source; **433** Reuters/CORBIS; **434** C. Sherburne/PhotoLink/Getty Images; **435** (l) MICHAEL S. YAMASHITA/National Geographic Image Collection; **438** (t to b) Digital Vision Ltd./SuperStock; (2)Masterfile; (3)Creatas/PictureQuest; (4)C. Sherburne/PhotoLink/Getty Images; **439** (t)C. Sherburne/PhotoLink/Getty Images; (b)MICHAEL S. YAMASHITA/National Geographic Image Collection; **440** (2,4)Macmillan/McGraw-Hill; (others)Hutchings Photography/Digital Light Source; **441** Hutchings Photography/Digital Light Source; **442** (t) PhotoLink/Getty Images; (b)C. Sherburne/PhotoLink/Getty Images; **444** supershoot/Alamy; **445** Daniel H. Bailey/Alamy; **448–449** George Frey/Getty Images; **450** Peter de Clercq/Alamy; **451** Hutchings Photography/Digital Light Source; **452** (l)Jan Tadeusz/Alamy; (r)matthias engelien/Alamy; **454** (l)WIN-Initiative/Getty Images; (c)MIMOTITO/Getty Images; (r) age fotostock/SuperStock; **456** (t)WIN-Initiative/Getty Images; (b)Jan Tadeusz/Alamy; **457** (t)NASA/Jeff Schmaltz, MODIS Land Rapid Response Team, (b)Jocelyn Augustino/FEMA; **458** Kyle Niemi/U.S. Coast Guard via Getty Images; **459** Hutchings Photography/Digital Light Source; **461** Hutchings Photography/Digital Light Source; **464** (l)Amazon-Images/Alamy; (c)Roger Coulam/Alamy; (r)mediacolor's/Alamy; **465** Eric Nguyen/CORBIS; **466** StockTrek/Getty Images; **467** AP Photo/Dick Blume, Syracuse Newspapers; **468** Eric Nguyen/CORBIS; **470** Gene Rhoden/Visuals Unlimited; **471** Hutchings Photography/Digital Light Source; **472** (l,r) National Oceanic and Atmospheric Administration (NOAA); **474** Dennis MacDonald/Alamy; **475** (t)National Oceanic and Atmospheric Administration (NOAA); (b)Dennis MacDonald/Alamy; **476** (t to b)Aaron Haupt; (others)Hutchings Photography/Digital Light Source; **478** (t)Peter de Clercq/Alamy; (c)AP Photo/Dick Blume, Syracuse Newspapers; (b)matthias engelien/Alamy; **481** George Frey/Getty Images; **484–485** Ashley Cooper/CORBIS; **486** Hemis.fr/SuperStock; **490** (tl)Rolf Hicker/age fotostock; (tc) Brand X Pictures/PunchStock; (tr)Jeremy Woodhouse/Getty Images; (bl) Andoni Canela/age fotostock; (br)Steve Cole/Getty Images; **492** age fotostock/SuperStock; **493** (t)Rolf Hicker/age fotostock; (b)Digital Vision/Getty Images; **494** (5)Macmillan/McGraw-Hill; (others)Hutchings Photography/Digital Light Source; **495** J. A. Kraulis/Masterfile; **496** (t) Hutchings Photography/Digital Light Source; (b)Nick Cobbing/Alamy; **502** Nick Cobbing/Alamy; **503** (t,b)American Museum of Natural History; **504** Ashley Cooper/Alamy; **505** Hutchings Photography/Digital Light Source; **507** Chris Cheadle/Getty Images; **508** Steve McCutcheon/Visuals Unlimited, Inc.; **510** Bruce Harber/age fotostock; **511** (t)Chris Cheadle/Getty Images; (b)Bruce Harber/age fotostock; **512** (t to b,2,3,5,7)Macmillan/McGraw-Hill; (r,4,6)Hutchings Photography/Digital Light Source; **514** (t)Andoni Canela/age fotostock; (c)J. A. Kraulis/Masterfile; (b)Steve McCutcheon/Visuals Unlimited, Inc.; **517** Ashley Cooper/CORBIS; **SR-00–SR-01** Gallo Images-Neil Overy/Getty Images; **SR-2** Hutchings Photography/Digital Light Source; **SR-6** Michell D. Bridwell/PhotoEdit; **SR-7** (t)The McGraw-Hill Companies, (b)Dominic Oldershaw; **SR-8** StudiOhio; **SR-9** Timothy Fuller; **SR-10** Aaron Haupt; **SR-12** KS Studios; **SR-13 SR-47** Matt Meadows; **SR-48** Stephen Durr, (c)NIBSC/Photo Researchers, Inc., (r)Science VU/Drs. D.T. John & T.B. Cole/Visuals Unlimited, Inc.; **SR-49** (t)Mark Steinmetz, (r)Andrew Syred/Science Photo Library/Photo Researchers, (br)Rich Brommer; **SR-50** David Fleetham/Visuals Unlimited/Getty Images, (l)Lynn Keddie/Photolibrary, (tr) G.R. Roberts; **SR-51** Gallo Images/CORBIS.

PERIODIC TABLE OF THE ELEMENTS

Legend:
- Gas
- Liquid
- Solid
- Synthetic

Element box example: Hydrogen — Element; 1 — Atomic number; H — Symbol; 1.01 — Atomic mass; State of matter indicator.

A column in the periodic table is called a **group**.

A row in the periodic table is called a **period**.

Period	1	2	3	4	5	6	7	8	9
1	Hydrogen 1 H 1.01 (Gas)								
2	Lithium 3 Li 6.94	Beryllium 4 Be 9.01							
3	Sodium 11 Na 22.99	Magnesium 12 Mg 24.31							
4	Potassium 19 K 39.10	Calcium 20 Ca 40.08	Scandium 21 Sc 44.96	Titanium 22 Ti 47.87	Vanadium 23 V 50.94	Chromium 24 Cr 52.00	Manganese 25 Mn 54.94	Iron 26 Fe 55.85	Cobalt 27 Co 58.93
5	Rubidium 37 Rb 85.47	Strontium 38 Sr 87.62	Yttrium 39 Y 88.91	Zirconium 40 Zr 91.22	Niobium 41 Nb 92.91	Molybdenum 42 Mo 95.96	Technetium 43 Tc (98) (Synthetic)	Ruthenium 44 Ru 101.07	Rhodium 45 Rh 102.91
6	Cesium 55 Cs 132.91	Barium 56 Ba 137.33	Lanthanum 57 La 138.91	Hafnium 72 Hf 178.49	Tantalum 73 Ta 180.95	Tungsten 74 W 183.84	Rhenium 75 Re 186.21	Osmium 76 Os 190.23	Iridium 77 Ir 192.22
7	Francium 87 Fr (223)	Radium 88 Ra (226)	Actinium 89 Ac (227)	Rutherfordium 104 Rf (267) (Synthetic)	Dubnium 105 Db (268) (Synthetic)	Seaborgium 106 Sg (271) (Synthetic)	Bohrium 107 Bh (272) (Synthetic)	Hassium 108 Hs (270) (Synthetic)	Meitnerium 109 Mt (276) (Synthetic)

The number in parentheses is the mass number of the longest lived isotope for that element.

Lanthanide series: Cerium 58 Ce 140.12 | Praseodymium 59 Pr 140.91 | Neodymium 60 Nd 144.24 | Promethium 61 Pm (145) (Synthetic) | Samarium 62 Sm 150.36 | Europium 63 Eu 151.96

Actinide series: Thorium 90 Th 232.04 | Protactinium 91 Pa 231.04 | Uranium 92 U 238.03 | Neptunium 93 Np (237) (Synthetic) | Plutonium 94 Pu (244) (Synthetic) | Americium 95 Am (243) (Synthetic)